Klaus I. Baumann • Zdenek Halata • Ingrid Moll (Eds.)

The Merkel Cell/Structure-Development-Function-Cancerogenesis

AF785994

Springer-Verlag Berlin Heidelberg GmbH

K. I. Baumann • Z. Halata • I. Moll (Eds.)

The Merkel Cell

Structure-Development-Function-Cancerogenesis

With 94 Figures, 23 in Color, and 12 Tables

Springer

Professor Dr. med. Klaus I. Baumann
Professor Dr. med. Zdenek Halata

Institute of Functional Morphology

Professor Dr. med. Ingrid Moll

Department of Dermatology and Venerology

University of Hamburg
UKE, Martinistrasse 52
20246 Hamburg
Germany

Proceedings of the International Merkel Cell Symposium, Hamburg 2002

ISBN 978-3-642-05574-4

Library of Congress Cataloging-in-Publication Data
International Merkel Cell Symposium (4th : 2002 Hamburg, Germany)
The Merkel Cell : structure development, function cancerogenesis / K. Baumann, Z. Halata, I. Moll (eds.).
p. cm.
"Proceedings of the International Merkel Cell Symposium, Hamburg 2002."
Includes bibliographical references and index.
ISBN 978-3-642-05574-4 ISBN 978-3-662-10358-6 (eBook)
DOI 10.1007/978-3-662-10358-6
1. Merkel cells--Congresses. 2. Merkel cell carcinoma--Congresses. I. Baumann, K., 1945- II. Halata, Zdenek, 1939- III. Moll, I., 1951-IV. Title.

This work is subject to copyright. All rights reserved, whether the whole or part of the material is concerned, specifically the rights of translation, reprinting, reuse of illustrations, recitation, broadcasting, reproduction on microfilm or in any other way, and storage in data banks. Duplication of this publication or parts thereof is permitted only under the provisions of the German Copyright Law of September 9, 1965, in its current version, and permission for use must always be obtained from Springer-Verlag Berlin Heidelberg GmbH. Violations are liable for prosecution under the German Copyright Law.

http://www.springer.de
© Springer-Verlag Berlin Heidelberg 2003
Originally published by Springer-Verlag Berlin Heidelberg New York in 2003
Softcover reprint of the hardcover 1st edition 2003

The use of general descriptive names, registered names, trademarks, etc. in this publication does not imply, even in the absence of a specific statement, that such names are exempt from the relevant protective laws and regulations and therefore free for general use.

Cover design: Design & Production GmbH, Heidelberg
Typesetting: Camera-ready by Klaus Baumann<
39/3150WI- 5 4 3 2 1 0 - Printed on acid-free paper

Preface

This book contains papers presented as talks or posters during the International Merkel Cell Symposium – Hamburg 2002. Since its first description by Friedrich Sigmund Merkel in 1875, the cell named after him has fascinated researchers from different backgrounds throughout the world. For a very long time, this cell has succeeded in fooling researchers and keeping many of its secrets. This has resulted in controversies in the literature concerning the origin of Merkel cells, their function in mechanoreceptors as well as the question whether different types of Merkel cells exist.

Merkel had no doubt that these cells were involved in the perception of touch. This notion was confirmed by the description of Haarscheiben (hair disks) by Pinkus in 1905. With the introduction of electron microscopy in the 1960s, a new wave of investigations into Merkel cells began. Iggo's group was the first to combine electron microscopy and electrophysiology, demonstrating a unique pattern of action potentials in response to mechanical stimuli and introducing the name slowly adapting type I (SA I) mechanoreceptor. The discovery of various neuropeptides in the granules of Merkel cells led in the 1970s to the assumption of neuroendocrine functions of these cells. At about the same time, a highly malignant type of skin cancer with cells resembling Merkel cells was first described by Toker's group and called Merkel cell carcinoma. These discoveries are reflected in the widespread scientific interest in Merkel cells, ranging from anatomists, embryologists, neurobiologists and physiologists to clinical dermatologists. This community of Merkel cell researchers provided a solid base for several International Merkel Cell Symposia held so far.

The first two symposia were held in Heidelberg in 1994 and 1997, while a third was held in Tokyo in 1999. Thus, this book gives an account of the discussions during the Fourth International Merkel Cell Symposium. It covers the areas of morphology, development, mechanoreceptor function and Merkel cell carcinoma and provides the current state of knowledge and thoughts for future investigations in these areas. Participants of this symposium and contributors to this book are based in ten countries throughout the world. On behalf of all participants, we would like to express our gratitude to the Deutsche Forschungsgemeinschaft for generous support of the symposium and this book.

It is planned to hold the Fifth International Merkel Cell Symposium again in Hamburg in 2005 which will be chaired by Ingrid Moll. We expect most participants of this symposium to return to Hamburg in 2005, and also hope to see many new faces in the growing scientific community interested in Merkel cells.

Hamburg, 2003 Klaus I. Baumann, Zdenek Halata, Ingrid Moll

International Merkel Cell Symposium, July 6 - 7, 2002 in Hamburg

Contents

Morphology

Development

Merkel Cell Mechanoreceptors

Merkel Cell Carcinoma

List of Contributors - List of Participants

Baumann, Klaus I. Institute of Functional Morphology, University of Hamburg
UKE, Martinistr. 52, D-20246 Hamburg, Germany
E-mail: kbaumann@uke.uni-hamburg.de

Beiras-Fernandez, Andres
Dept. of Morphol. Science & Histology
University of Santiago de Compostela
E-15705 Santiago, Spain
E-mail: abcabezon@yahoo.com

Beiras-Iglesias, Andres
Dept. of Morphol. Science & Histology
University of Santiago de Compostela
E-15705 Santiago, Spain
E-mail: dphistme@usc.es

Brandner, Johanna Dept. of Dermatology, University of Hamburg
UKE, Martinistr. 52, D-20246Hamburg

Cahusac, Peter M.B. Dept. of Psychology, University of Stirling
Sirling FK9 4LA, UK
E-mail: p.m.b.cahusac@stir.ac.uk

Cook, Tony Dept. of Dermatology, University of Qeensland
Queensland Radium Institute Laboratory
Herston, Queensland, Australia 4029

Dehnhardt, Guido Dept. of Zoology & Neurobiology, University of Bochum
Bldg. ND6/Rm. 33, D-44780 Bochum, Germany
E-mail: dehnhardt@neurobiologie.ruhr-uni-bochum.de

Ebara, Satomi Department of Anatomy
Meiji University of Oriental Medicine
Kyoto 629-0392, Japan
E-mail: s_ebara@mx.meiji-u.ac.jp

Espinosa de los Monteros, Antonio
Department of Comparative Pathology
Veterinary Faculty, University of Las Palmas de
Gran Canaria, Las Palmas de Gran Canaria, Spain
E-mail: espinosa@cicei.ulpgc.es

Fukuda, Jun
Dept. of Physiology, Mol & Cell Biol. Lab
National Defense Medical College of Japan
3-2 Namiki, Tokorozawa, Japan 359-0042
E-mail: jfukuda@cc.ndmc.ac.jp

Garcia-Caballero, Thomas
Dept. of Morphol. Science & Histology
University of Santiago de Compostela
E-15705 Santiago, Spain
E-mail: cm-tgc@usc.es

Germain, Lucie
LOEX, Hop St. Sacrement du CHA
1050 Chemin St. Foy
Quebec City, PQ G1S 4L8, Canada
E-mail: Lucie.Germain@chg.ulaval.ca

Grim, Milos
Institute of Anatomy, 1st Medical Faculty
Charles University of Prague, U Nemocnice 3
CZ-12800 Praha, Czech Republic
E-mail: Milos.Grim@lf1.cuni.cz

Haass, Nikolas
Dept. of Dermatology, University of Hamburg
UKE, Martinistr. 52, D-20246Hamburg
E-mail: haass@uke.uni-hamburg.de

Halata, Zdenek
Institute of Functional Morphology, University of Hamburg
UKE, Martinistr. 52, D-20246 Hamburg, Germany
E-mail: halata@uke.uni-hamburg.de

Houdek, Pia
Dept. of Dermatology, University of Hamburg
UKE, Martinistr. 52, D-20246Hamburg

Ishizaki, Ken
Dept. of Complete Denture Prosthodontics
Tokyo Dental College
1-2-2, Masago, Mihamaku, Chiba 261-8502, JAPAN
E-mail: ishizaki@tdc.ac.jp

Kief, Sabine
Dept. of Dermatology, University of Hamburg
UKE, Martinistr. 52, D-20246Hamburg

Leonard, Helen J.
Dept. of Dermatology, University of Qeensland
Queensland Radium Institute Laboratory
Herston, Queensland , Australia 4029
E-mail: helenL@qimr.edu.au

Magnani, Dino	Dept. of Dermatology, University of Hamburg UKE, Martinistr. 52, D-20246Hamburg E-mail: magnani@uke.uni-hamburg.de
Masaki, Yoshinori	Dept. of Physiology, Mol & Cell Biol. Lab National Defense Medical College of Japan 3-2 Namiki, Tokorozawa, Japan 359-0042
Moll, Ingrid	Dept. of Dermatology, University of Hamburg UKE, Martinistr. 52, D-20246Hamburg E-mail: moll@uke.uni-hamburg.de
Moll, Roland	Dept. of Pathology, Philipps University Marburg Baldinger Str., D-35033 Marburg, Germany E-mail: mollr@mailer.uni-marburg.de
Nagai, Ko	Department of Biomolecular Sensing (Dermatology) Yamaguchi University School of Medicine Minami- Kogushi 1-1-1 Ube 755-8505 Japan
Narisawa, Yutaka	Div. of Dermatology, Dept. of Internal Med. Saga Medical School, Nabeshima 5-1-1 Saga 849-8501, Japan E-mail: narisawa@post.sag-med.ac.jp
Peters, Eva	Psychoneuroimmunology Research Group Department of in Internal Medicine, Psychosomatics Charité, Campus Virchow Clinic, Biomedical Research Center R. 2.0534 Augustenburger Platz 1, D-13353 Berlin, Germany Email: eva.peters@charite.de
Ramirez Rivero, Gustavo	Department of Comparative Pathology Veterinary Faculty, University of Las Palmas de Gran Canaria, Las Palmas de Gran Canaria, Spain E-mail: gustavo@cicei.ulpgc.es
Rice, Frank L.	Dept. of Anatomy & Neurobiol., Albany Medical College 47 New Scotland Ave, Albany, NY 12208, USA E-mail: RiceF@mail.amc.edu
Sames, Klaus	Institute of Neuroanatomy, University of Hamburg UKE, Martinistr. 52, D-20246 Hamburg, Germany E-mail: sames@uke.uni-hamburg.de

Schulz, Tilman
Institute of Pathology, Klinikum Nürnberg
Prof. Ernst Nathan Str. 1, D-90419 Nürnberg
E-mail: t.schulz@klinikum-nuernberg.de

Schumacher, Udo
Institute of Neuroanatomy, University of Hamburg
UKE, Martinistr. 52, D-20246 Hamburg, Germany
E-mail: uschumac@uke.uni-hamburg.de

Senok, Solomon S.
Dept. of Physiology, College of Medicine and Medical Sciences, Arabian Gulf University
PO Box 22979, Manama, Bahrain
E-mail: solomonss@agu.edu.bh

Sieber-Blum, Maya
Dept. of Cell Biology, Neurobiology and Anatomy
Medical College of Wisconsin
8701 Watertown Plank Rd, Milwaukee, WI 53226, USA
E-mail: sieberbl@mcw.edu

Suzuki, Yuta
Dept. of Cr&Br Prosthodontics, Tokyo Dental College
1-2-2, Masago, Mihamaku, Chiba 261-8502, Japan
E-mail: yusuzki@tbc.ac.jp

Tachibana, Tamiko
Dept. of Oral Anatomy, Iwate Medical University
Chuo-Dori 1-3-27, Morioka 020-8505, Japan
E-mail: tatamiko@iwate-med.ac.jp

Tadokoro, Osamu
Dept. of Oral Anatomy, Matsumoto Dental University
Shiojiri, Nagano-ken 399-0781, Japan
E-mail: osamut@po.mdu.ac.jp

Takahashi-Igawanaga, Hiromi
Dept. of Anatomy, Hokkaido University, Grad. School of Medicine, Kita-15, Nishi-7, Sapporo 060-8638, Japan
E-mail: hiromiva@tara.med.hokudai.ac.jp

Tazaki, Masakazu
Dept. of Physiology, Tokyo Dental College
1-2-2, Masago, Mihamaku, Chiba 261-8502, Japan
E-mail: mtazaki@tbc.ac.jp

van Gele, Mireille
Center for Medical Genetics, University Hospital Gent
De Pinclaan 185, 1K5, B-9000 Gent, Belgium
E-mail: mireille.vangele@rug.ac.be

Vesper, Meikel
Dept. of Orofacial Surgery, University of Hamburg
UKE, Martinistr. 52, D-20246 Hamburg, Germany
E-mail: m.vesper@uke.uni-hamburg.de

Morphology

Merkel Nerve Endings Functioning as Mechanoreceptors in Vertebrates

Zdenek Halata[1], Klaus I. Baumann[1], Milos Grim[2]

[1]Dept. of Functional Anatomy, University of Hamburg, UKE Martinistr. 52, 20246 Hamburg, Germany

[2]Institute of Anatomy, First Faculty of Medicine, Charles University Prague, 12800 Prague, Czech Republic

Summary

This study focused on those types of mammalian nerve endings originally described by Merkel in 1875. In all mammals, Merkel cells are found in the basal layer of the epithelium usually in close contact with nerve terminals. The Merkel cell is always positioned between the direction of the mechanical stimulus and the nerve terminal. Any similar cell without nerve contact may have neuroendocrine, but not mechanoreceptor functions. Thus, such cells should not be called Merkel cells, as there is no evidence for any relationship with those structures originally described by Merkel.

Introduction

In 1875, Friedrich S. Merkel first described "Tastzellen" (touch cells) and "Tastkörperchen" (touch corpuscles) in avian dermis and oral mucosa as well as in mammalian epidermis (Merkel 1875). In all cases, the Merkel cells were closely associated with nerve terminals. The name indicates that Merkel was convinced to look at mechanoreceptors.

Results and Discussion

Merkel Nerve Endings in the Mammalian Glabrous Skin

In contrast to the location of Merkel cells and Grandry corpuscles in the connective tissue below the epidermis in birds, Merkel cells in the mammalian glabrous skin are always found in the basal layer of the epidermis (Fig. 1). There are two types of glabrous skin in mammals. The first type (pegged skin) has solid epidermal pegs of different size anchoring the epidermis in the dermis. A typical example is found in the pig snout and planum nasale of mole and cat. The dermis contains blood sinus separating the epithelial pegs. In the basal layer of the epidermis at the base of these pegs are clusters of up to 40 Merkel cells. Smaller pegs may only have 6 Merkel cells. All Merkel cells are in contact with terminals of myelinated axons (3–5 µm) losing their axon on entering the epidermis and branching into several discoid terminals with large numbers of mitochondria and electron microscopically empty vesicles. While in the nose of the mole only one axon innervates the Merkel cells of an epithelial peg, there are several axons supplying one epithelial peg in the pig snout.

Merkel cells are oval in shape with the longitudinal axis parallel to the basal lamina (Fig. 2). The nucleus is large and lobulated. The cytoplasm contains bundles of intermediary filaments and osmiophilic granules of about 60 nm adjacent to the nerve terminal. Some regions of the membrane between Merkel cell and nerve terminal have synapse-like structures, while on the opposite site cytoplasmic processes extend into and between keratinocytes linked with desmosomes.

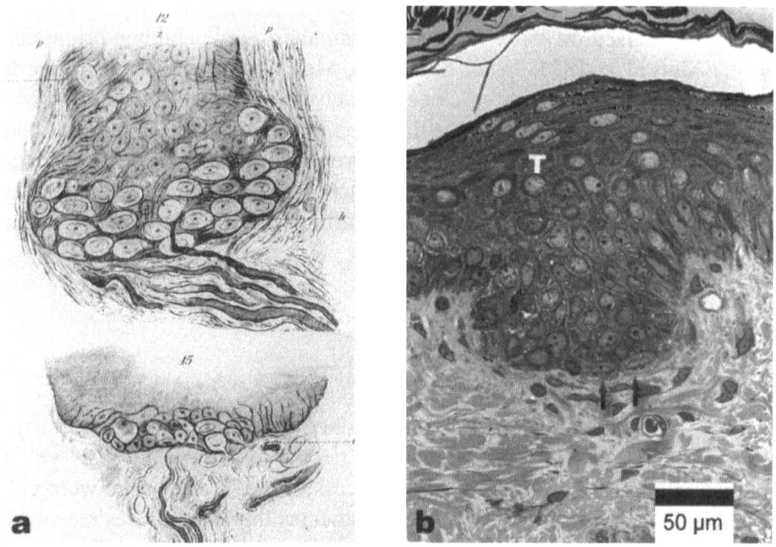

Fig. 1. a Original drawings from Merkel's publication (1875) illustrating "touch corpuscles" in mammalian glabrous skin. **b** Semithin section of a "Tastscheibe" (*T*) with Merkel nerve endings (*arrows*) in epidermal rete peg from the eyelid of a rhesus monkey

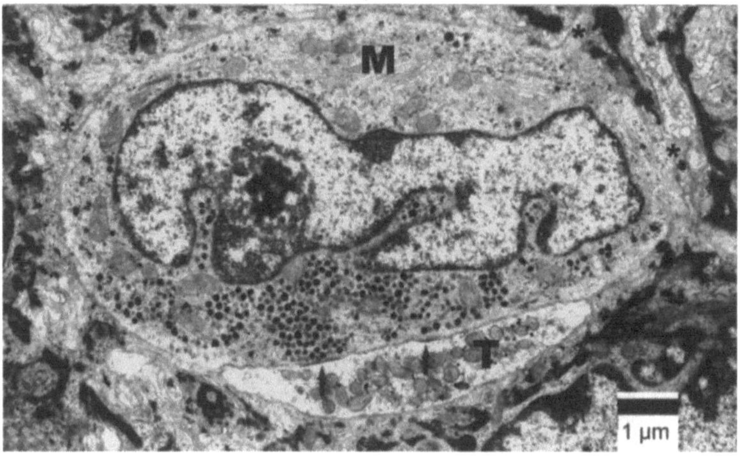

Fig. 2. High power electron microscopy of a Merkel cell (*M*) with nerve terminal (*T*) from the rete peg skin of the cat nose. Synapse-like contacts can be seen between Merkel cell and nerve terminal (*arrows*)

The second type of skin (ridged skin) is typically found in tips of fingers and toes of primates and marsupials. Clusters of up to ten Merkel nerve endings are found at the base of the epidermal ridges near the ducts of sweat glands. Moreover, in this location the long axis of the oval Merkel cells runs parallel to the surface with the nerve terminals below the Merkel cells towards the basal lamina.

Merkel Nerve Endings in Hairy Skin and Whiskers

Between hair follicles of the hairy skin, the epidermis forms epithelial pegs of different size and density, often referred to as touch disks ("Tastscheiben"). These are basically similar to the epithelial pegs in glabrous skin (see above), containing variable numbers of Merkel nerve endings, depending on the size of the touch dome. In addition, Merkel cells with a similar arrangement are found in the thickening of hair follicles below the sebaceous gland of guard and velus hairs.
Whiskers are large hair follicles embedded in blood sinus with strong sturdy hairs. Merkel nerve endings in connection with whiskers are found in two locations. A small number can be seen at the transition between hair follicle and surrounding epidermis. Much larger numbers (up to 2,000) are located in the basal layer of the epithelium of the hair follicle in the thickened part below the sebaceous gland. This area is surrounded by the ring sinus. Merkel cells are arranged oblique to the basal lamina (glassy membrane) sending cytoplasmic processes of up to 15 µm through the glassy membrane (Fig. 3). All of them are in contact with discoid nerve terminals on the opposite side. The part of Merkel cells facing nerve terminals contains the typical dense core (osmiophilic) granules. After losing their myelin sheath and Schwann cells, the axons branch intensively supplying up to 50 Merkel cells. A more detailed review of morphology, development and physiology of Merkel cells is given in Halata et al. (2003).

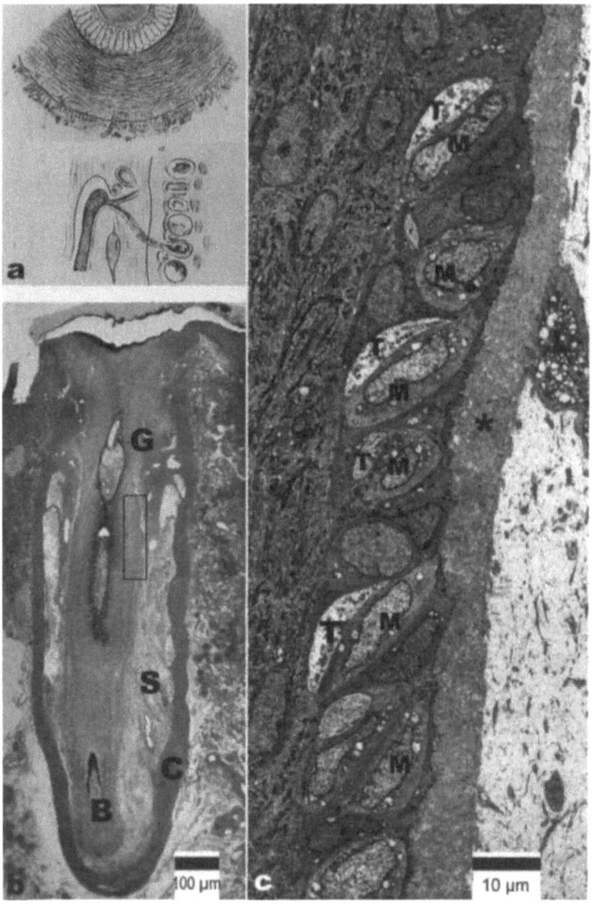

Fig. 3. a Original drawing from Merkel's publication (1875) illustrating "touch cells" in pig sinus hairs. **b** Semithin longitudinal section through a rhesus monkey sinus hair from the upper lip in low magnification showing a sebaceous gland *(G)*, the cavernous blood sinus (*S*), the connective tissue capsule of the sinus (*C*) and the hair bulb (*B*). The *square* indicates the area shown enlarged in **c**. **c** Ultrathin longitudinal section through the thickened portion of the sinus hair follicle below the sebaceous gland. Merkel cells (*M*) and nerve terminals (*T*) are arranged obliquely to the glassy membrane (*) like the scales of a pine cone, with the Merkel cells always directed towards the glassy membrane. A lanceolate nerve terminal (*L*) in close contact with the glassy membrane (*) is seen on the outer side

References

Halata Z, Gim M, Baumann KI (2003) Friedrich Sigmund Merkel and his "Merkel Cell", morphology, development and physiology – review and new results. Anat Rec 271:225–239

Merkel F (1875) Tastzellen and Tastkoerperchen bei den Hausthieren und beim Menschen. Arch Mikr Anat 11:636–652

The Three-Dimensional Microanatomy of Merkel Cells

Hiromi Takahashi-Iwanaga

Department of Anatomy, Hokkaido University Graduate School of Medicine, Kita-15, Nishi-7, Sapporo 060-8638, Japan

Summary

Merkel cells, mechanoreceptor cells in epithelial tissues, have generally been regarded as functioning to detect tissue deformations with their microvilli and release certain transmitters to their targets. Cutaneous Merkel cells in the sinus hair follicle are known to possess many secretory granules close to their accompanying axon endings, while mucosal Merkel cells in the palate are reported to extend paracrine processes to the basal lamina. A comparative observation of these two Merkel cell types in rats by a combination of scanning and transmission electron microscopy showed that cutaneous Merkel cells displayed a triangular body with a sharpened apex and a flattened base in the basal layer of the external root sheath. Distribution of microvilli was confined to the apical and basal aspects of the cells. The apical microvilli were tightly held by corresponding pits in overlying epithelial cells, while the basal ones deeply penetrated the basement membrane. On the other hand, mucosal Merkel cells in the basal layer of the palatal epithelium exhibited a round cell body, and extended membranous processes to the basal lamina. Microvilli were randomly dispersed along the entire cell surface. Some microvilli terminated as free ends, while others were stuck in adjacent epithelial cells. The different manners of support of the possible receptor sites indicate different properties of the two Merkel cell types in their response to mechanical stimuli.

Introduction

The touch cell of Merkel was originally described as a bubble-like cell associated with a nerve ending in the epidermis of mammals (Merkel 1875). Transmission electron microscopy (TEM) has characterized the cell by its spine-like microvilli radiating from the cell body and numerous secretory granules apposed to the nerve terminal (Cauna 1962; Munger 1965). These microvilli have generally been regarded as functioning to detect tissue deformations, and the granules to contain mediators of mechanoreception, since Iggo and Muir (1969) have shown that the Merkel cell–neurite complex operates as a slowly adapting mechanoreceptor by the electrophysiological recording of afferent fibers.

In their extensive TEM studies, Tachibana et al. (1997) demonstrated numerous variants of Merkel cells in the oral mucosa of rodents. The cells, designated here as mucosal Merkel cells as distinct from typical cutaneous Merkel cells, frequently failed to contact the nerve fibers, and extended dendritic processes filled with secretory granules to the epithelial basal lamina. These cells have been assumed to exert certain paracrine functions, possibly nursing epithelial cells, or guiding nerve fibers to their target.

The present study comparatively analyzes the three-dimensional histotopography of the two Merkel cell types with special reference to their microvilli and paracrine processes in order to define their morphological and functional diversity. Observations are reported on cutaneous Merkel cells in sinus hair follicles (Patrizi and Munger 1966; Halata 1993) and mucosal ones in the posterior palate of rats. The specimens were observed both by scanning electron microscopy (SEM) after exposure of the cells by NaOH maceration, and by TEM according to a conventional method.

Materials and Methods

Detailed procedures for the preparation of Merkel cell specimens have been reported elsewhere (Takahashi-Iwanaga and Abe 2001). Briefly, 6-week-old male rats of the Wistar strain were anesthetized with sodium pentobarbital, and perfusion-fixed with a mixture of 2.5% glutaraldehyde and 0.5% paraformaldehyde buffered at pH 7.4 with 0.1 M cacodylate. The palatal mucosa in the intermolar and postrugal regions and the sinus hair follicles in mystacial pads were excised and immersed in the same fixative overnight. The fixed tissues were cut into small pieces and processed for SEM and TEM observation.

Specimens to be examined by SEM were macerated with 6 N NaOH at 60 °C for 15 min (Takahashi-Iwanaga and Fujita 1986). After maceration, the tissue pieces were postfixed with 1% tannic acid for 1 h, followed by 1% OsO_4 for 1 h. The osmicated specimens were dehydrated through a graded series of ethanol, transferred to isoamyl acetate, and critical-point dried with liquid CO_2. The dried specimens were coated with osmium, and examined in a Hitachi H-4500 scanning electron microscope.

For TEM observation, the aldehyde-fixed specimens were immersed in 1% OsO_4 for 2 h at 4 °C, dehydrated through a series of ethanol, and embedded in Epon-812. Ultrathin sections were examined in a Hitachi H-7100 transmission electron microscope after double staining with uranyl acetate and lead citrate, or after triple staining with uranyl acetate, lead citrate, and tannic acid.

Results

The NaOH maceration of the SEM specimens effectively hydrolized the basement membrane, and loosened cell junctions to expose the basal and lateral aspects of epithelial tissues. Merkel cells were identified by their spine-like microvilli dispersed on the cell surface both in the sinus hair follicle and in the palatal mucosa (Fig. 1). By TEM, the cells were seen to enclose numerous dense cored granules in the cytoplasm, and frequently formed desmosome junctions with adjoining epithelial cells (Fig. 2).

Cutaneous Merkel Cells in the Sinus Hair Follicle

Numerous Merkel cells occurred in the basal layer of the external root sheath at the level of the ring sinus as reported previously (Patrizi and Munger 1966; Halata 1993). At this part, the basal layer consisted of small epithelial cells elongated at right angles to the hair axis, while its overlying layer of the root sheath comprised large squamous cells. The elongated basal cells and the cutaneous Merkel cells were roughly similar in size and shape, except that the latter displayed a triangular cross section with a narrow apical ridge and a flattened base in contrast to the cuboidal appearance of the former. The basal cells hung over the apical ridge of the Merkel cells, and extended flap-like projections beneath their cell base to entirely envelope the cell body (Fig. 1a, b).

The apical ridge and the basal surface of the cutaneous Merkel cells were equipped with numerous microvilli measuring 2–3 μm in length, while the lateral sides of the cells were absolutely smooth. The apical microvilli projected into corresponding pits in the overlying squamous cells through gaps between basal cells. The basal microvilli of the Merkel cells protruded from the basal layer, and penetrated the thick basement membrane as corroborated by TEM (Fig. 2a). Lateral sides of the Merkel cells were frequently attached to discoid endings of axons with secretory granules of the former accumulated in the contacting area.

Mucosal Merkel Cells in the Palate

Merkel cells clustered in the basal layer of the epithelial ridges of the posterior palatal mucosa, in accord with previous reports (Tachibana et al. 1997). The epithelial ridges were represented by a stack of polyhedral cells increasing in size toward the surface of the mucosa. The mucosal Merkel cells revealed a small, round cell body and large membranous processes extended along intercellular spaces in the epithelium (Fig. 1b, c). Both the cell body and the processes were attached to the basal lamina through gaps between basal cells. Each Merkel cell body was surrounded by three to five basal cells on the lateral aspect, and contacted by a larger epithelial cell on the apical aspect.

Microvilli of the mucosal Merkel cells were about 2 μm long, and randomly dispersed on the cell body and on the membranous processes. Some microvilli

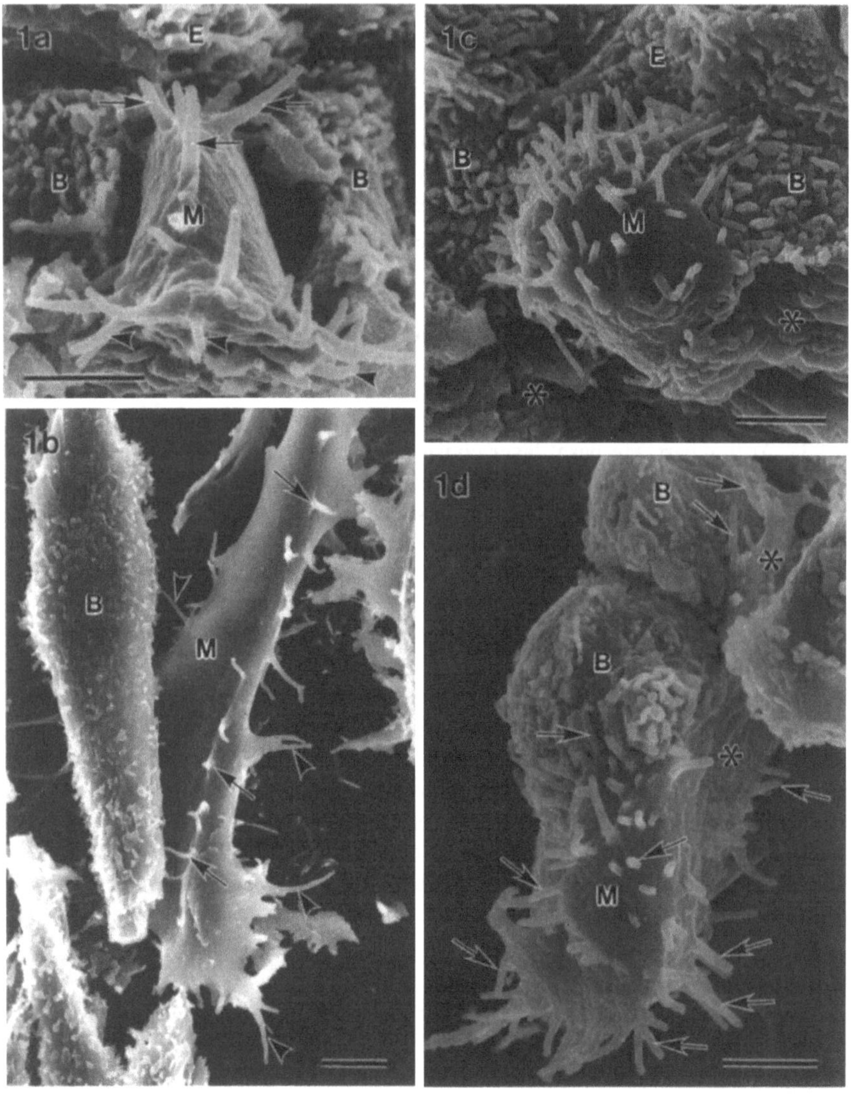

Fig. 1a–d. Scanning electron micrographs of rat Merkel cells. **a** Cross section of a sinus hair follicle. A Merkel cell (*M*) in the basal layer displays a triangular body with apical (*arrows*) and basal (*arrowheads*) microvilli. *B* Basal cells, *E* squamous epithelial cell overlying the basal layer. **b** Cells in a basal layer of a sinus hair follicle. A Merkel cell (*M*) issues microvilli from the apical ridge (*arrows*) and the flattened base (*arrowheads*). *B* Basal cell. **c** Cross section of palatal epithelium. A Merkel cell (*M*) radiates microvilli to adjoining basal cells (*B*), as well as to an overlying epithelial cell (*E*). *Asterisks* Basal surface of the epithelium. **d** Cells in a basal layer of palatal epithelium. A Merkel cell (*M*) embraces two basal cells (*B*) with its membranous process (*asterisks*). Note numerous microvilli (*arrows*) dispersed along the surface of the Merkel cell. *Bars* 3 μm

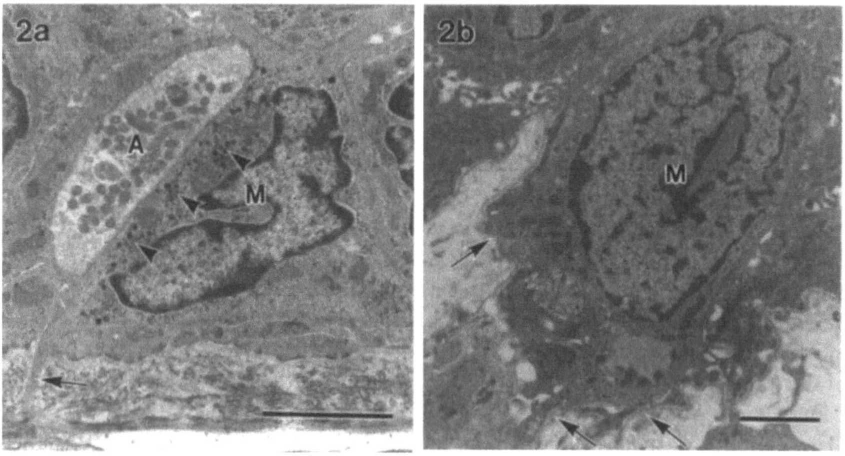

Fig. 2a, b. Rat Merkel cells presented by transmission electron microscopy. **a** Basal layer of a sinus hair follicle. A Merkel cell (*M*) extends a spine-like projection into basement membrane (*arrow*). Note numerous dense cored granules (*arrowheads*) apposed to an axon ending (*A*). **b** Basal layer of palatal epithelium. A Merkel cell body (*M*) containing fine granules is attached to the basal lamina (*arrows*). *Bars* 2 µm

penetrated adjoining epithelial cells while others terminated as free ends in intercellular spaces. Synaptic contacts were only occasional in the mucosal Merkel cells. Secretory granules occurred in the entire cytoplasm of the cells (Fig. 2b).

Discussion

The cutaneous Merkel cells in the sinus hair follicle and the mucosal ones in the posterior palate of rats exhibited their characteristic features as reported previously by TEM (Patrizi and Munger 1966; Halata 1993; Tachibana et al. 1997): the former frequently displayed synaptic contacts with axon endings, while the latter attached paracrine processes to the basal aspect of the epithelium. Using immunohistochemistry with cell markers, Tachibana et al. (1997) described the processes of the mucosal Merkel cells as pursuing long dendritic courses among epithelial cells, suggesting their paracrine effects on these cells. In the present SEM observation, the dendritic Merkel processes revealed their membranous shapes, apposing larger and more numerous secretory surfaces to their possible targets.

Merkel cells have generally been believed to randomly radiate their microvilli (Iggo and Muir 1969). However, the combined SEM and TEM observations of the present study have shown that the cutaneous Merkel cell in the sinus hair follicle and the mucosal one in the posterior palate differ from each other in the distribution of their microvilli on the cell surface as well as in their topographical relationships with surrounding tissues. In the former cell, microvilli were restricted to the apical and basal aspects. The apical and basal microvilli were tightly held by overlying epithelial cells and the basement membrane, respectively, while the cell body was shielded from these abutments by adjoining basal cells. In the latter mucosal Merkel cell, on the other hand, microvilli were randomly dispersed along the entire cell surface. These microvilli, as well as the main portion of the cell, were arbitrary connected with their surrounding structures.

There is accumulated evidence showing that Merkel cells respond to mechanical stretching of the cell membrane by an elevation of the intracellular Ca^{2+} concentration, which triggers granule exocytosis in various secretory cells (Chan et al. 1996; Tazaki and Suzuki 1998). It is reasonable to expect that attenuated microvilli of the Merkel cells are the most susceptible sites for such stimuli. The well-organized tissue support of the cutaneous Merkel cell suggests that microvilli of the cell are specifically displaced by a certain type of tissue deformation in favor of the potential role of the cell as a mechanoreceptor. On the other hand, the random anchoring of the mucosal Merkel cell indicates that various kinds of tissue deformations nonspecifically bend its microvilli, ensuring the incessant release of secretory granules for the maintenance of epithelial cells and development of nerve fibers.

References

Cauna N (1962) Functional significance of the submicroscopical, histochemical and microscopical organization of the cutaneous receptor organs. Anat Anz 111:181–197

Chan E, Yung W-H, Baumann KI (1996) Cytoplasmic Ca^{2+} concentrations in intact Merkel cells of an isolated, functioning rat sinus hair preparation. Exp Brain Res 108:357–366

Halata Z (1993) Sensory innervation of the hairy skin (light- and electron-microscopic study). J Invest Dermatol 101:75S–81S

Iggo A, Muir AR (1969) The structure and function of a slowly adapting touch corpuscle in hairy skin. J Physiol 200:763–796

Merkel F (1875) Tastzellen und Tastkörperchen bei den Hausthieren und beim Menschen. Arch Mikrosk Anat 11:636–652

Munger BL (1965) The intraepidermal innervation of the snout skin of the opossum. A light and electron microscopic study with observations on the nature of Merkel's "Tastzellen". J Cell Biol 26:79–97

Patrizi G, Munger BL (1966) The ultrastructure and innervation of rat vibrissae. J Comp Neurol 126:423–436

Tachibana T, Yamamoto H, Takahashi N, Kamegai T, Schibanai S, Iseki H, Nawa T (1997) Polymorphism of Merkel cells in the rodent palatine mucosa: immunohistochemical and ultrastructural studies. Arch Histol Cytol 60:379–389

Takahashi-Iwanaga H, Abe K (2001) Scanning electron microscopic observation of Merkel cells in the lamprey epidermis. Acta Anat Nippon 76:375–380

Takahashi-Iwanaga H, Fujita T (1986) Application of an NaOH maceration method to a scanning electron microscopic observation of Ito cells in the rat liver. Arch Histol Jpn 49:349–357

Tazaki M, Suzuki T (1998) Calcium inflow of hamster Merkel cells in response to hyposmotic stimulation indicate a stretch activated ion channel. Neurosci Lett 243:69–72

Staining of Living Merkel Cells with FM Dyes

Jun Fukuda, Yoshinori Masaki and Hisako Ishimine

Laboratory of Molecular and Cellular Physiology, Department of Physiology, National Defense Medical College of Japan, 3-2 Namiki, Tokorozawa, Saitama 359-8513, Japan

Summary

Live Merkel cells are known to incorporate a fluorescent dye, quinacrine (3,3′,4′,5,7-pentahydroxyflavone). Quinacrine fluorescence in the cells is, however, quickly lost and quinacrine-stained Merkel cells become difficult to identify in tissue culture. To find dyes that stay in the cells for a long period of time, we tested many fluorescent dyes and found that FM1-43 (N-(3-triethylammoniumpropyl)-4-(4-(dibutylamino)-styryl)pyridinium dibromide) is a useful marker for live Merkel cells. In the rat footpad skin, FM1-43 was revealed to stain most of the live Merkel cells that were already stained with quinacrine. Merkel cells in sinus hair follicles were also stained with FM1-43 dye. The fluorescence intensity of the FM dye was stronger than that of quinacrine, and the shape of the cells was more distinct in the FM1-43-stained cells. We thus conclude that the FM dye is a powerful tool for tracing live Merkel cells in in vitro experiments. Moreover, the finding that Merkel cells incorporate the FM dye suggests that vesicles in Merkel cells are likely to represent recycling in a manner similar to those in neurons and secretory cells.

Introduction

Live Merkel cells in the skin and hair are known to incorporate quinacrine (Crowe and Whitear 1978; Nurse and Diamond 1984) which elicits fluorescent light under UV-illumination. The method of marking live Merkel cells with quinacrine has been utilized to differentiate the cells from other epidermal cells in in vivo and in vitro experiments (Nurse et al. 1983, 1984; Nurse and Diamond 1984). This dye is, however, quickly lost from the Merkel cells. To find fluorescent dyes that distinguish live Merkel cells from other epidermal cells, we tested many fluorescent dyes and found that FM1-43 dye is a powerful marker for identifying live Merkel cells (Fig. 1).

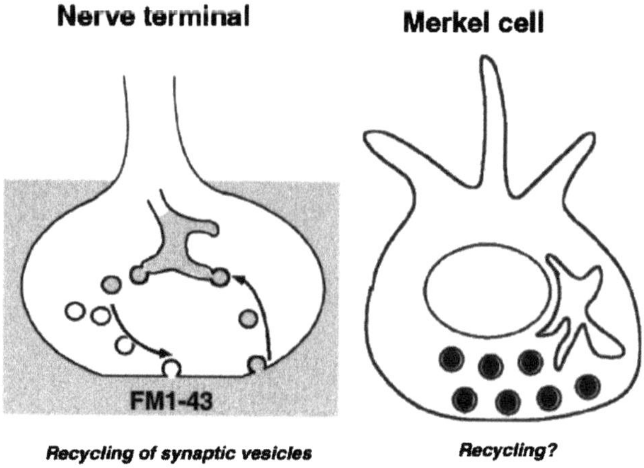

Fig. 1. Schematic illustrations of the mechanisms of staining cells with the FM dye. *Left* FM1-43 dye added to the extracellular space is incorporated in the nerve terminal during recycling of synaptic vesicles. *Right* Vesicles in Merkel cells may represent recycling similar to those in the synaptic terminal. In fact, it is demonstrated in the present study that Merkel cells incorporate FM1-43 dye

Materials and Methods

Merkel Cells in Skin of Footpad. A skin sample was dissected from the footpad of rats (7–14 days old) and was cut into pieces of approximately 3×3 mm. These pieces were first incubated in Dulbecco/F-12 medium containing 20 ng/ml quinacrine at 37 °C for more than 2 h in a CO_2 incubator. In order to remove connective tissue, these pieces were incubated in ice-cold Dulbecco/F-12 medium containing trypsin (0.04%) and pancreatin (36 mg/ml) for 30 min. After bathing the preparations in Dulbecco/F-12 medium containing 1% protease inhibitors (Sigma, USA) and 10% FCS (Gibco, USA), the connective tissue was carefully removed with forceps under a dissection binocular microscope. Merkel cells thus exposed on the surface of the preparation were examined under a fluorescence microscope. The same preparations that were examined under the microscope were then incubated with 10 μg/ml FM1-43 dissolved in high-K (20 mM) Hank's solution for 30 min at 37 °C. The preparations were rinsed with Dulbecco/F-12 medium containing 10% FSC for more than 2 h to remove nonspecifically bound FM dyes.

Merkel Cells in Sinus Hair Follicles. Sinus hair follicles were dissected from a whisker pad of a young rat, and their capsules were removed under a binocular microscope. To stain Merkel cells with FM dyes, preparations were incubated in FM-dye-containing high-K Hank's solution for 30 min at 37 °C. In order to remove connective tissue, these preparations were incubated with trypsin/pancreatin and the connective tissue was removed.

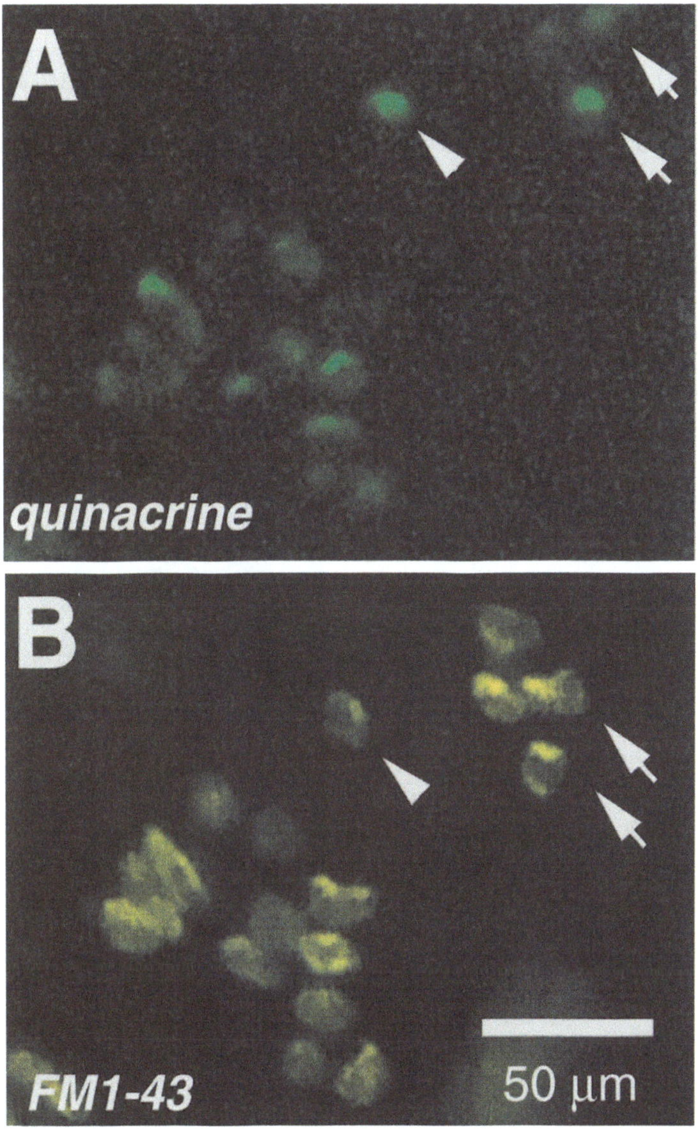

Fig. 2A, B. Merkel cells in rat footpad skin stained with quinacrine and FM dye. The footpad skin was first exposed to quinacrine (**A**) and then to FM1-43 (**B**). **A** Computer-enhanced fluorescent picture of quinacrine-stained Merkel cells. **B** The same skin preparation shown in **A** was stained with FM1-43. A group of cells indicated by *arrows* and a cell indicated by an *arrowhead* changed their position during the perfusion of the preparation. *Scale bar* 50 μm

Results

Merkel Cells in Footpad Skin Stained with Quinacrine and FM1-43. Figure 2 demonstrates that live Merkel cells in the footpad skin are stained with both quinacrine and FM1-43. We first identified Merkel cells in the skin preparation by staining with quinacrine (Fig. 2A), and these quinacrine-stained cells were then stained with FM1-43 (Fig. 2B). A group of small cells of approximately 10–20 μm are seen to generate fluorescence of quinacrine (Fig. 2A). According to the size and shape of the cells, they were concluded to be Merkel cells. The same preparation as that shown in Fig. 2A was then incubated in a high-K (20 mM) Hank's medium containing 10 μM FM1-43 for 1 h. After rinsing the nonspecifically bound FM dye by perfusion for 2 h, the same preparation was again viewed under a fluorescence microscope (Fig. 2B). It was observed that most quinacrine-positive cells shown in Fig. 2A are also stained with FM1-43. The fluorescence intensity of FM1-43 was stronger than that of quinacrine and the contour of the cells was more distinct in the FM-dye-stained skin preparation.

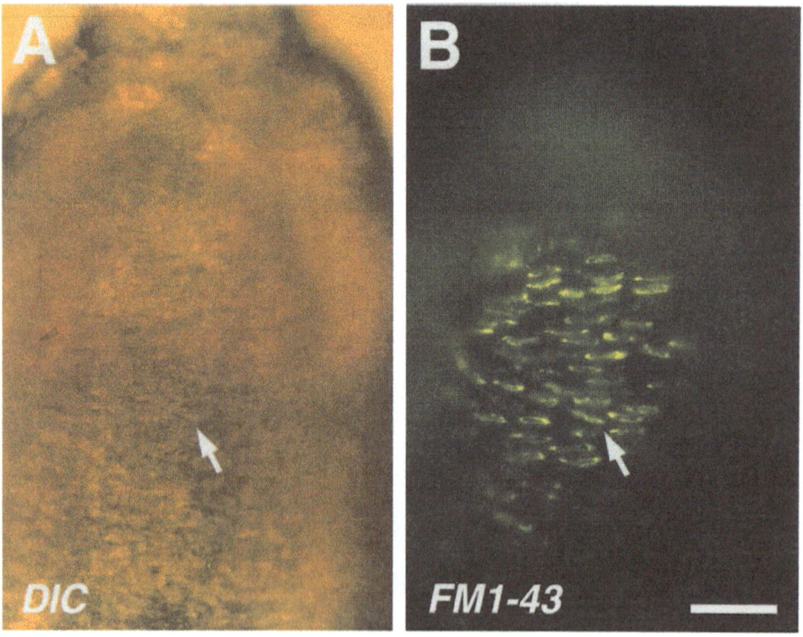

Fig. 3A, B. Merkel cells in sinus hair follicles stained with FM1-43 dye. **A** DIC picture of superior enlargement of a hair follicle. Connective tissue covering the superior enlargement was removed by protease treatment (see Materials and Methods). **B** a fluorescent picture of the same visual field in **A**. FM-dye-stained cells (*arrow*) were seen on the surface of the preparation

Merkel Cells in Sinus Hair Follicles. Merkel cells of sinus hair follicles from a whisker pad were stained with FM1-43 or quinacrine (Fig. 3). To obtain clear images, the thin connective tissue surrounding the superior enlargement (Andres 1966) of the hair follicle was removed by protease treatment (Fig. 3A). When viewed under a fluorescence microscope (Fig. 3B), FM-dye-stained cells (arrows in Fig. 3B) were seen lined along the surface of the preparation. It was noted that the long axes of these spindle-shaped cells were approximately vertical to the hair (arrowhead in Fig. 3A) of the follicle.

Discussion

The present study shows that live Merkel cells in the skin and hair follicles are stained with FM dyes. Accordingly, we conclude that FM dyes are convenient probes for differentiating the cells from other epidermal cells. A small number of quinacrine-positive cells were not stained with FM dyes. This was likely due to the instability of Merkel cells on the surface of the protease-treated preparations.

Vital staining of Merkel cells with FM dyes has advantages in several respects as compared with that with quinacrine. First, the fluorescence intensity of FM dyes is stronger than that of quinacrine. Indeed, in tissue culture preparations, FM-dye-positive Merkel cells were more easily identified than quinacrine-positive cells under a fluorescence microscope. Second, the contour of Merkel cells in the skin and hair follicles is more distinct with FM-dye-staining than with quinacrine-staining. The clear image of FM-dye-positive Merkel cells is useful to monitor the change in the shape of the cells. Third, compared with quinacrine, FM dyes remain in the cells for a long period of time. Merkel cells that were once stained with the dyes could be monitored for more than a week in our culture experiments (Fukuda et al., in press).

FM dyes are known to be incorporated in synaptic vesicles in neurons (Betz et al. 1996; Cochilla et al 1999; Cousin 1999; Sudof 2000; Ryan 2001) and secretory cells (Angleson et al. 1999). Thus, they have been frequently utilized as tools for studying the mechanisms of turnover of synaptic vesicles. The present finding indicates that Merkel cells are different from other epidermal cells in terms of their ability to incorporate FM dyes. Since Merkel cells are known to contain synaptic vesicles, FM dyes may be useful tools for studying the mechanisms of turnover of these vesicles in Merkel cells.

References

Andres KH (1966) Über die Feinstruktur der Rezeptoren an Sinushaaren. Z Zellforsch 75:339–365

Angleson JK, Cochilla AJ, Kilic G, Nussinovitch I, Betz WJ (1999) Regulation of dense core release from neuroendocrine cells revealed by imaging single exocytic events. Nat Neurosci 2:440–446

Betz WJ, Mao F, Smith CB (1996) Imaging exocytosis and endocytosis. Curr Opin Neurobiol 6:365–371

Cochilla AJ, Angleson JK, Betz WJ (1999) Monitoring secretory membrane with FM1-43 fluorescence. Annu Rev Neurosci 22:1–10

Cousin MA (1999) Mechanisms of synaptic vesicle recycling illuminated by fluorescent dyes. J Neurochem 73:2227–2239

Crowe R, Whitear M (1978) Quinacrine fluorescence of Merkel cells in *Xenopus laevis*. Cell Tissue Res 190:273–283

Fukuda J, Ishimine H, Masaki Y (2003) Long-term staining of live Merkel cells with FM dyes. Cell Tissue Res (in press)

Nurse CA, Diamond J (1984) A fluorescent microscopic study of the development of rat touch domes and their Merkel cells. Neurosci 11:509–520

Nurse CA, Mearow KM, Holmes M, Visheau B, Diamond J (1983) Merkel cell distribution in the epidermis as determined by quinacrine fluorescence. Cell Tissue Res 228:511–524

Nurse CA, Macintyre L, Diamond J (1984) A quantitative study of the time course of the reduction in Merkel cell number within denervated rat touch domes. Neuroscience 11:521–53

Ryan TA (2001) Presynaptic imaging techniques. Curr Opin Neurobiol 11:544–549

Sudhof TC (2000) The synaptic vesicle cycle revisited. Neuron 28:317–320

Distribution of Merkel Cells in Normal Skin and Mucosae in Dogs

A. Espinosa de los Monteros[1], F. Rodríguez[1], J. Martín de las Mulas[2], A. Lara[1], G.A. Ramírez[1]

[1]Department of Comparative Pathology, Veterinary Faculty, University of Las Palmas de Gran Canaria, Las Palmas de Gran Canaria, Spain
[2]Department of Comparative Pathology, Veterinary Faculty, University of Córdoba, Córdoba, Spain

Summary

Merkel cells (MCs) are widely distributed in human skin, localized in a high proportion in the epidermis of hairy and glabrous skin in sites of high tactile sensitivity. In other mammals, snouts, touch domes, sinus hair follicles, lips and hard palate are rich areas of MCs. The aim of this study was to determine the normal distribution of MCs in the dog. Tissue samples from several locations were obtained from six adult dogs of different age, sex and breed, fixed in 10% buffered formalin, embedded in paraffin wax and cut into 3 μm tissue sections. The ABC method was used with different poly- and monoclonal antibodies tested (anti-Ck20, anti-Cks 8 and 18, anti-CgA, anti-NSE, anti-synaptophysin, anti-NF). MCs were observed in several sites of the skin and the oral, genital and nasal mucosae, as well as in different types of hair follicles. Face skin, digital pads, oral mucosae and vibrissae appear as high-density areas of MCs in dogs, in which they are frequently disposed in groups of three to ten cells. In many locations, the cells showed close relation with terminal axons. These locations are in accordance with studies in other mammals reported in the literature.

Introduction

Merkel cells (MCs) were first described by F. Merkel in 1875 who called these cells "Tastzellen" (touch cells; Merkel 1875). They are found in the basal layer of the epidermis and epithelium of mucosae in mammals, where they are in contact with epidermal cells, as well as in hair follicles. MCs have been related to mechanical perception (Tachibana 1995). Moreover, other possible functions such as introduction of growing nerve fibres to their destinations, participation of the morphogenesis of skin appendages (Tachibana 1995) and induction of the development of taste buds (Toyoshima et al. 1999) have been reported.

In humans, MCs are located in the epidermis of hairy and glabrous skin, predominantly in sites of high tactile sensitivity such as the plantar and digital skin (Hartschuh et al. 1989), and in hair follicles (Uchigasaki et al. 2000). Snouts, touch domes, sinus hair follicles, lips and hard palate are rich areas of MCs in other mammals including cats, pigs, guinea pigs, rats and monkeys (Tachibana 1995).

The purpose of the study was to determine the normal location of Merkel cells in the canine skin and mucosae and the areas in which this cell population reach the highest proportion.

Materials and Methods

Skin and mucosae samples from several locations (Table 1; Figs. 1, 2) were obtained from six dogs of different breeds (boxer, mixed-breed dog (2), poodle (2) and cocker spaniel), age (between 2 and 4 years) and sex (three males and three females). Tissue samples were fixed in 10% buffered formalin for 24 h, embedded in paraffin and cut at 3 µm. The different sections were deparaffined, rehydrated in graded alcohols, and incubated with 3% hydrogen peroxidase in methanol for 30 min to block endogenous peroxidase activity. Heat-induced antigen retrieval (water bath at 95–100 °C) with 10 mM citrate buffer (pH 6.0, 5–6 min) was used. After cooling, slides were covered with 10% normal goat serum in PBS for monoclonal antibodies and in 10% normal swine serum for polyclonal antibodies for 30 min before incubation with the primary antibodies for 18 h at 4 °C. The different primary antibodies employed were cytokeratins 8, 18 and 20, neuron-specific enolase, chromogranin A, synaptophysin, and neurofilaments. The avidin-biotin peroxidase complex (ABC) method was applied (Vector Laboratories, Burlingame, CA) and the chromogen used was 3-amino-9-ethyl-carbazole. The slides were counterstained with Mayer's hematoxylin.

The immunostainings were evaluated separately and results were based upon the consensus of at least two observers. The number of immunoreactive cells was evaluated on a 3-point scale on 1.5-cm-long tissue sections (+ 1–6 cells, ++ 7–12 cells, +++ >12 cells).

Results and Discussion

Merkel cells (MCs) were distributed in different proportions according to the corporal area analyzed (Table 1), showing differences between mucosae, skin and hair follicles.

Table 1. Locations and distribution of Merkel cells in skin, mucosae and hair follicles

Skin		MCs	Mucosa		MCs
Head	Nasolabiale planum	+++	Oral	Hard palate	+++
	Nasal apex and planum nasale	+++		Soft palate	++
	Cheeks (skin)	+++		Cheeks (mucosa)	+++
	Lower lip (skin)	+++		Gums	+++
	Upper lip (skin)	+++		Tongue	+
	Eyelids (skin)	+		Lower lip (mucosa)	+++
	Pinna (external side)	-		Upper lip (mucosa)	++
	Pinna (internal side)	+	Genital	Clitoris	+
Trunk	Dorsal region	+		Vulva (mucosa)	-
	Lumbar region	+		Vagina	-
	Flanks	-		Prepuce (mucosa)	-
	Abdomen and inguinal region	+		Glans penis	+
	Axilla region and chest	+	Ocular	Eyelid conjunctiva	-
	Nipples	-		Ocular conjunctiva	-
	Vulva (skin)	+		Third eyelid	-
	Scrotum	-	Anal	Anocutaneous junction (mucosa)	-
	Prepuce (skin)	+	Nasal	Nasal vestibule	+
	Anocutaneous junction (skin)	+			
Limbs	Fore limb	+			
	Hind limb	+		**Hair Follicles**	**MCs**
	Fore limb digital pads	+++		Tylotrichs	+++
	Hind limb digital pads	+++		Vibrissae	+++
	Interdigital skin	+		Normal	++

-: absent; +: 1-6 cells; ++: 7-12 cells; +++: > 12 cells (in 1,5 cm tissue sections).

Mucosae

Oral mucosa was the area in which the highest population of MCs was found, mainly in the hard and soft palate, internal mucosa of cheeks and gums. They were distributed alone or forming groups of 3–14 cells, with a round to oval morphology. The larger proportion of MCs was located in basal layers of the epithelium, although occasionally a few cells were observed in more superficial layers. This spherical cell population was the only observed and dendritic MCs observed by

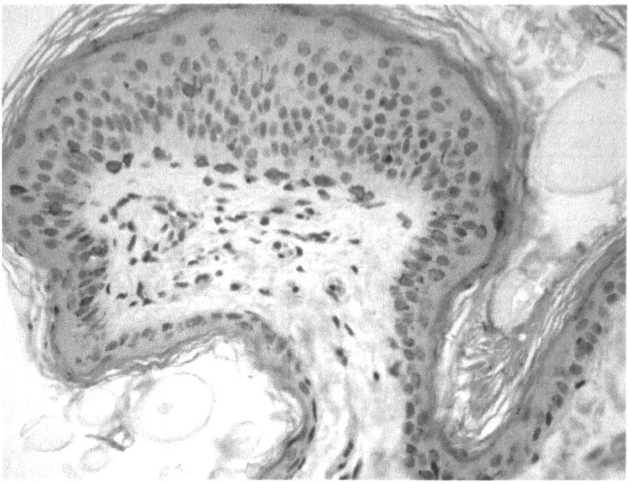

Fig. 1. Dog, interdigital skin. Group of Merkel cells in the basal layer of a Haarscheibe. Note the disposition of the cells with the long axis parallel to the skin surface. Cytokeratin 20 antibody, ABC method, AEC chromogen, Mayer's hematoxylin counterstaining; ×400

other authors were not noted (Tachibana and Nawa 2000). Evidence of close contact between spherical MCs and terminal nerves were observed with neuron-specific enolase (Fig. 2). In nasal and genital mucosa, a very small number of MCs were seen. They had oval to elongate morphology and were always alone and in the basal layer of the epithelium. MCs were not observed in other mucosal location.

Skin

MCs were dispersed in the skin of the animal, specifically located in characteristic small thickenings of the epidermis irregularly distributed called "Haarscheiben" or "hair discs", first described by F. Pinkus in 1902. In these structures, MCs had an oval to round morphology and tended to form groups of three to ten cells (Fig. 1), rarely were they found as lonely cells. The immunostained cells lay in a plane parallel to the basement membrane and thus at right angles to the columnar epithelial cells. Sometimes, immunostained nerve fibres (neurofilaments and neuron-specific enolase) were clearly visible in the subcutaneous area, penetrating the basement membranes and in close association with the basal part of immunostained cells (Merkel cell–neurite complex). This fact has been observed in the skin of cats and rats (Gu et al. 1981).

The areas with high density in MCs were skin of the cheeks (mainly in the vicinity of the vibrissae), nasal apex and digital pads (where the arrangement of MCs is similar to that in mucosae), planum nasale, lips and nasolabiale planum (Table 1).

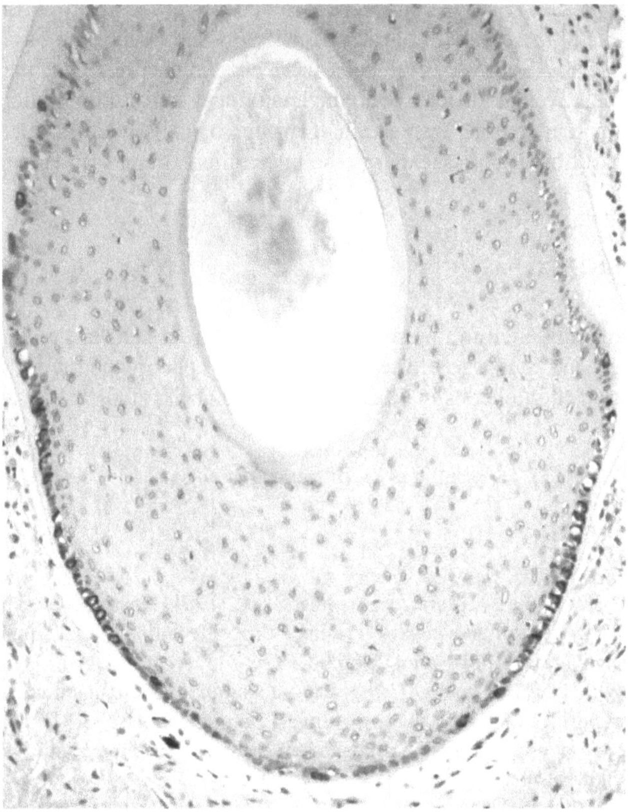

Fig. 2. Dog cheek. Vibrissae showing numerous immunoreactive Merkel cells in the outer root sheath in a transversal section. NCL-5D3 antibody, ABC method, AEC chromogen, Mayer's hematoxylin counterstaining. ×400

Hair Follicles

MCs were observed in normal or guard hair follicles, tylotrich hair follicles and vibrissae.

- Normal hair follicles. The cells were distributed alone and scattered among the epithelial cells of the bulge area. There were no significant differences in the numbers of MCs present in hair follicles of the different regions studied. They were always scarce (one to three MCs per follicle) with the exception of the dorsal region skin, in which the number of MCs was slightly increased.
- Tylotrich hair follicles. MCs were found in higher numbers and also scattered, being more numerous in the skin of the face mainly in the adjacent areas to the lips. Examinations of the differences between the average number of Merkel cells per hair follicle in humans revealed that there were more in facial follicles than in those elsewhere (Uchigasaki et al. 2000). In tylotrich hair follicles associated with Haarscheiben in the trunk skin, MCs were not as numerous and located in the upper infundibulum.

- Vibrissae. Positive cells were located in a very high proportion (>20 MCs per vibrissae) in the enlarged portion of the vibrissae, within the basal layer of the outer root sheath. Some of these cells were always vacuolated in the longitudinal sections. There was no difference in the anatomical distribution pattern among different anatomical regions tested. Small groups of cells were frequently found in the outlet part of the vibrissae and in the interfollicular areas (in Haarscheiben).

Conclusion

The results obtained in this study show that Merkel cells are widely distributed in hairy skin and mucosae of the dog and are in concordance with reports where the location of Merkel cells in other mammals has been studied.

References

Gu J, Polack JM, Tapia FJ, Marangos PJ, Pearse AG (1981) Neuron-specific enolase in the Merkel cells of mammalian skin. The use of specific antibody as a simple and reliable histologic marker. Am J Pathol 104:63–68

Hartschuh W, Weihe E, Egner U (1989) Chromogranin A in the mammalian Merkel cell: cellular and subcellular distribution. J Invest Dermatol 93:641–648

Merkel F (1875) Tastzellen und Tastköperchen bei den Hausthieren und beim Menschen. Arch Mikr Anat 11:636–652

Pinkus F (1905) Ueber Hautsinnesorgane neben dem menschlichen Haar (Haarscheiben) und ihre vergleichend-anatomische Bedeutung. Arch Mikr Anat 65:121–179

Tachibana T (1995) The Merkel cell: recent findings and unresolved problems. Arch Histol Cytol 58:379–396

Tachibana T, Nawa T (2000) Morphological and developmental characteristics of oral mucosal Merkel cells. In: Suzuki H, Ono T (eds) Merkel cells, Merkel cell carcinoma and neurobiology of the skin. Elsevier, Amsterdam, pp 11–21

Toyoshima K, Seta Y, Toyono T, Takeda S (1999) Merkel cells are responsible for the initiation of taste organ morphogenesis in the frog. J Comp Neurol 406:129–140

Uchigasaki S, Suzuki H, Takano K, Inoue K (2000) Characteristics of Merkel cells in hair follicles in the facial skin. In: Suzuki H, Ono T (eds) Merkel cells, Merkel cell carcinoma and neurobiology of the skin. Elsevier, Amsterdam, pp 115–119

Merkel Cells in Human Transplanted Flaps

M. Vesper[1], P. Houdek[2], I. Moll[2]

[1]Abteilung für Zahn-, Mund-, Kiefer- und Gesichtschirurgie, Universitätsklinikum Hamburg-Eppendorf, Martinistr. 52, 20246 Hamburg, Germany
[2]Klinik und Poliklinik für Dermatologie und Venerologie, Universitätsklinikum Hamburg-Eppendorf, Martinistr. 52, 20246 Hamburg, Germany

Summary

New procedures in the field of reconstructive surgery place more emphasis on the reconstruction of form and function. An important aspect is the touch perception of the transplanted tissue surface. The results of this study, involving the free radial forearm flap and the latissimus dorsi flap, allow the assumption that a nerve reconstruction is not necessary, because good reinnervation in most parts of the flaps can be observed. Merkel cells can be found within the hairy skin of most body regions. For the detection of Merkel cells CK 20 antibodies were used. Seven days after operation as well as after 2 and 4 months postoperative, CK 20-positive cells were found in the transplant. These results correlated with good recovery of touch perception.

Introduction

In continuous implementation of new procedures in the field of reconstructive surgery, primarily involving the head and neck region, more and more emphasis is laid on the reconstruction of form and function (Boyd et al. 1994; Katou et al. 1995; Baumann et al. 1996). An important aspect is the touch perception of the transplanted tissue surface.

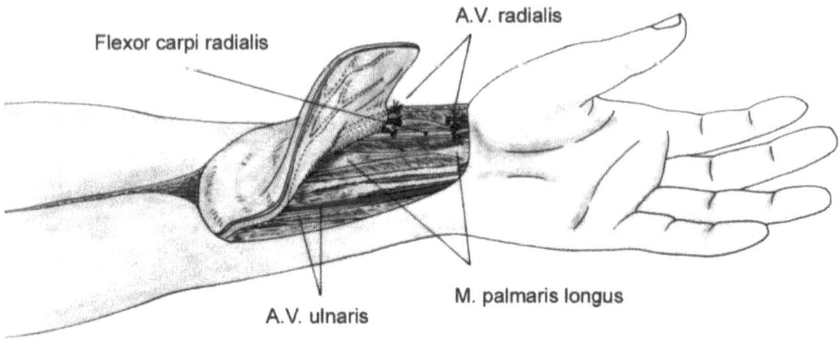

Fig. 1. Schematic drawing of the forearm flap

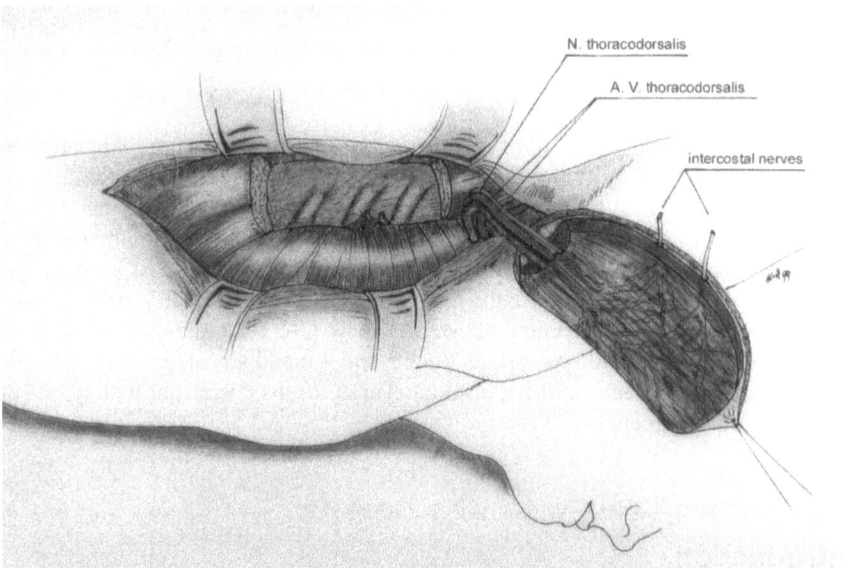

Fig. 2. Schematic drawing of the musculus latissimus dorsi flap

Materials and Methods

The results of our clinical study, involving the free radial forearm flap (19 patients, 10 females, 9 males, 11–71 years, mean 47.1 years; Fig. 1) and the musculus latissimus dorsi flap (62 patients, 30 females, 32 males, 3–85 years, mean 52.1 years; Fig. 2) support the assumption that nerve reconstruction is not necessary, because good reinnervation in most parts of the flaps can be observed. As a clinical examination method, Semmes-Weinstein monofilaments were used.

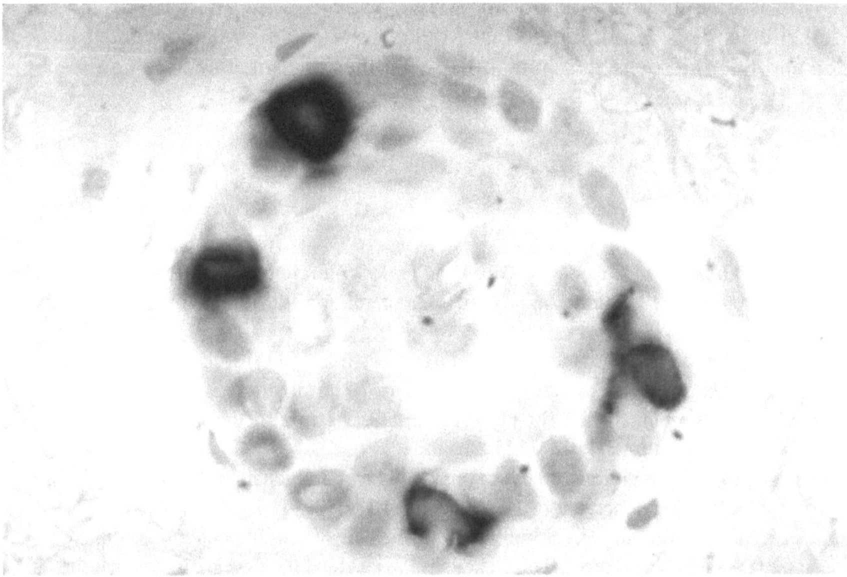

Fig. 3. Merkel cells 4 months after a free musculus latissimus dorsi flap stained for CK 20

These are often applied for the determination of the sensory recovery (Levin et al. 1978; Costas et al. 1994; Posnick and Grossmann 2000). Every flap was split into five sectors for the examination, one central and four peripheral (9–12, 12–3, 3–6 and 6–9 o'clock).

Results and Discussion

In the radial forearm group, we could detect 84.2% positive, in the musculus latissimus dorsi group 69.7% positive sectors with Semmes-Weinstein monofilaments. The examination time in the first group was between 2 and 135 months (mean 20.3 months). The examination time in the second group was between 1 and 127 months (mean 21.7 months).

Merkel cells can be found within the hairy skin of most body regions. We know that after nerve reconstruction domes with Merkel cells degenerate, but not always disappear completely (English 1977; English et al. 1983, 1984, 1992).

They may act as target sites which attract regenerating type I nerve fibers (Horch 1984; He et al. 1999). Maybe they are involved in the formation of new dome structures after nerve regeneration.

For the detection of Merkel cells, CK 20 antibodies were used. Seven days after operation as well as after 2 and 4 months postoperative, CK 20 positive cells were found in the transplants (Fig. 3). These results correlated with good recovery of touch perception, examined with Semmes-Weinstein monofilaments.

This is a preliminary report about five patients who all reacted positive to Merkel cells. It is the first time that Merkel cells could be demonstrated in pedicled and free microvascular human flaps. Further investigations are now necessary.

References

Baumann I, Greschnik A, Bootz F, Kaiserling E (1996) Freie transplantierte, mikrovaskulär reanastomosierte Unterarmlappen zur Rekonstruktion in Mundhöhle und Oropharynx. HNO 44:616–623

Boyd B, Mulholland S, Gullane P, Rotstein L, Brown D (1994) Reinnervated lateral antebrachial neurosome flaps: are we making sense? Plast Reconst Surg 93:1350–1359

Costas PD, Heatley G, Seckel BR (1994) Normal sensation of the human face and neck. Plast Reconst Surg 93:1141–1145

English KB (1977) The ultrastructure of cutaneous type I mechanoreceptors (Haarscheiben) in cats following denervation. J Comp Neurol 172:137–164

English KB, van Norman DK, Horch K (1983) Effects of chronic denervation in type I cutaneous mechanoreceptors (Haarscheiben). Anat Rec 207:79–88

English KB, Horch K, van Norman DK, Burgess PR (1984) Neurothropic influences on cutaneous type I mechanoreceptors. In: Hamann W, Iggo A (eds) Sensory receptor mechanisms. World Scientific, Singapore, pp 36–45

English KB, Wang ZZ, Stayner N, Stensaas LJ, Martin H, Tuckett RP (1992) Serotonin-like immunoreactivity in Merkel cells and their afferent neurons in touch domes from the hairy skin of rats. Anat Rec 232:112–120

He L, Tuckett RP, English KB (1999) Chemosensitivity of the rat type I slowly adapting mechanoreceptor. Biol Signals Receptors 8:382–389

Horch K (1984) Specificity in regenerating cutaneous nerves. In: Hamann W, Iggo A (eds) Sensory receptor mechanisms. World Scientific, Singapore, pp 47–55

Katou F, Shirai N, Kamekura S, Ohki H, Motegi K, Andoh N, Date F, Nagura H (1995) Intraoral reconstruction with innervated forearm flap. Oral Surg Oral Med Oral Pathol Oral Radiol 80:638–644

Levin S, Pearsell G, Rudermann W (1978) Von Freys method of measuring pressure sensibility in the hand: an engineering analysis of the Weinstein-Semmes pressure aesthesiometer. J Hand Surg 3:211–216

Posnick JC, Grossmann JA (2000) Facial sensibility testing: a clinical update. Plast Reconst Surg 106:892–894

Merkel Cells in Malassez Epithelium

Osamu Tadokoro[1], Katsuhiro Inoue[1], Takeyasu Maeda[2], Yukishige Kozawa[3], Inger Hals Kvinnsland[4], Vaska Vandevska Radunovic[5]

[1]First Department of Oral Anatomy, Matsumoto Dental University, Nagano, Japan
[2]Division of Oral Anatomy, Department of Oral Biological Science, Niigata University Graduate School of Medical and Dental Sciences, Niigata, Japan
[3]Department of Anatomy, Nihon University, School of Dentistry at Matsudo, Chiba, Japan
[4]Department of Anatomy and Cell Biology, University of Bergen, Bergen, Norway
[5]Department of Orthodontics, University of Oslo, Oslo, Norway

Summary

Recent findings indicate that Malassez epithelium (ME) consists of a heterogeneous epithelial cell population. The aim of this study was to investigate the identity of these cells using light and electron microscopy, using protein gene product 9.5 (PGP-9.5). Positive PGP-9.5 immunoreactive (IR) products were recognized in the ground substance of the epithelial cells in both human and cat periodontal ligament (PDL). These immunopositive cells in the cat ME shared the ultrastructural features of Merkel cells (MCs). No neural elements were observed in the vicinity of the cat ME. However, a close relationship between PGP-9.5 IR nerves and immunopositive epithelial cells in the human ME was revealed. This ultrastructural evidence demonstrates the existence of non-innervated MCs in cat ME and suggests the existence of innervated MCs in human ME.

Introduction

The epithelial cells, known as ME, in the PDL are remnants of Hertwig's epithelial root sheath. They exist as strands close to the cementum surface after tooth eruption (Hamamoto et al. 1989). Morphological studies have shown that ME in humans and rats is composed of light and dark cells (Brice et al. 1991; Kaneko et al. 1999), and there are cytoskeletal similarities between the basal junctional epithelium and ME (Peters et al. 1995). These data imply a non-uniform, heterogeneous structure of ME which has been reported.

Recent study of cat PDL confirms the cellular diversity of ME. The cat ME has been demonstrated to contain cells immunoreactive to CGRP, VIP and SP (Kvinnsland et al. 2000). Furthermore, a double immunolabelling for PGP-9.5 and cytokeratin (CK) 20 of cells within ME and gingival epithelium gives strong support to the hypothesis that ME also includes Merkel-like cell properties (Tadokoro et al. 2002). The aim of this study was to investigate the ultrastructure of PGP-9.5 immunopositive cells in human and cat ME and gingival epithelium using light and electron microscopy.

Materials and Methods

Specimens from teeth and tooth-supporting tissues in human and cat were used in this study. Human PDL tissue including maxillary incisors was fixed in formalin, rinsed in phosphate buffer and decalcified in formic acid. The cats were anaesthetised with s Nembutal and perfused with paraformaldehyde. The jaws were excised, post-fixed and decalcified in EDTA. After demineralisation, the teeth were cut sagittally at 30 μm.

For immunohistochemical purposes, a standard procedure was used. A polyclonal antibody to PGP-9.5 was used and the antigen detection was determined with the ABC method. Final visualisation was performed by DAB, with or without nickel ammonium sulphate. Following confirmation of the DAB reaction, some cat sections were post-fixed in 1% OsO_4, dehydrated and then embedded in epoxy resin. Ultrathin sections were examined under a transmission electron microscope.

Results

The observation demonstrated the presence of PGP-9.5 IR cells and nerve fibres in both human and cat. Immunopositive cells were also detected in the ME and the apices of the rete pegs of the gingival epithelium. The majority of these cells were oval in shape and located in the vicinity of nerve fibres.

Some positive cells in the cat gingival epithelium contained PGP-9.5 IR products in the ground substance closer to the cell membrane. Their nuclei were irregular in shape, with secretory granules characteristic for MC. However, there were no nerve endings in their vicinity.

Epithelial cells in the cat PDL were connected by desmosome-like structures and surrounded with basal lamina (Fig. 1). The ground substance of these cells contained cored vesicles of approximately 100 nm and tonofilaments. These vesicles, appearing as secretory granules, were accumulated in finger-like projections protruding towards the neighbouring epithelial cells and PDL cells.

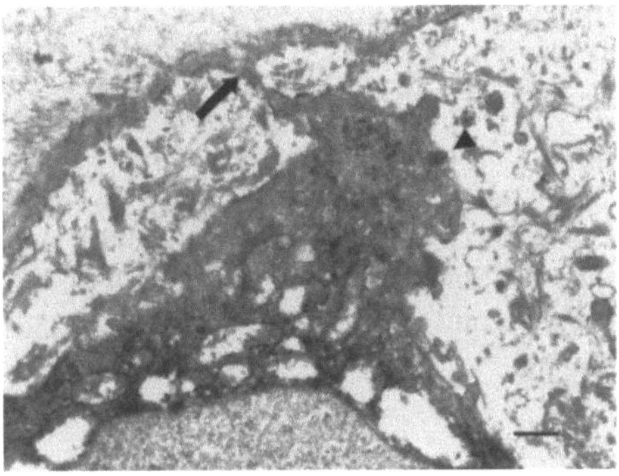

Fig. 1. PGP-9.5 IR cells in cat ME. Dense core vesicles (*arrowhead*) accumulate in the cytoplasmic process and finger-like projections protruding toward the PDL fibres (*arrow*) can be seen. *Scale bar* 500 nm

Discussion

The present study revealed the existence of PGP-9.5 IR cells in human and cat ME and gingival epithelium. The ultrastructural morphology of the immunopositive epithelial cells in the Malassez clusters lead us to the hypothesis that they represent Merkel-like cells.

Previously, we showed that neuroendocrine cells contain CGRP, SP and VIP in cat ME and gingival epithelium (Kvinnsland et al. 2000). Furthermore, single cells in cat ME and gingival epithelium show simultaneous immunolabelling for PGP-9.5 and CK 20 (Tadokoro et al. 2002), indicating that these neuroendocrine cells may be categorised in MCs. The PGP-9.5 IR cells shown in this study had large cored vesicles, characteristic for MCs. They were found in the cell body, as well as the cytoplasmic processes protruding towards the neighbouring cells. Taken together, the PGP-9.5 IR cells have possible neuromodulatory roles for ME in the cat PDL.

MCs can be classified into two types; innervated and non-innervated MCs (Tachibana et al. 1997). Under an electron microscope, the PGP-9.5 IR cells were non-innervated MCs. However, previous studies and this light and electron microscopic study have shown an intimate relation between nerve endings and ME (Kvinnsland et al. 2000; Tadokoro et al. 2002), indicating that both types are present in the ME of humans and cats.

The finger-like projections in the MCs in the cat ME resemble the axon terminals of periodontal Ruffini endings (Byers and Maeda 1997). The role of MCs still remains unclear (Tachibana 1995), but it is possible that they serve as mechanoreceptors.

The present results indicate that cat ME contains non-innervated MCs. Further investigations are needed to clarify the functional role of MCs in the periodontal ME.

Acknowledgements

We thank Drs. Hiroyuki Mishima and Kunihiro Suzuki, Dept of Anatomy, Nihon University School of Dentistry at Matsudo for their assistance and Drs. Mitsuyoshi Hagio and Yuichi Hidaka, Department of Veterinary Surgery, Miyazaki University for their supply of materials. The study was supported by grants from Sasakawa Grants for Science Fellows (SGSF2002), Matsumoto Dental University(1999, 2001) and grant-in-aid from the Ministry of Education, Science, Sport, and Culture, Japan.

References

Brice G, Sampson WJ, Sims MR (1991) An ultrastructural evaluation of the relationship between epithelial rests of Malassez and orthodontic root resorption and repair in man. Aust Orthod J 12:90–94

Byers MR, Maeda T (1997) Periodontal innervation: regional specializations, ultrastructure, cytochemistry and tissue interactions. Acta Med Dent Helv 2:116–133

Hamamoto Y, Nakajima T, Ozawa H (1989) Ultrastructural and histochemical study on the morphogenesis of epithelial rests of Malassez. Arch Histol Cytol 52:61–70

Kaneko H, Hashimoto S, Enokiya Y, Ogiuchi H, Shimono M (1999) Cell proliferation and death of Hertwig's epithelial root sheath in the rat. Cell Tissue Res 298:95–103

Kvinnsland IH, Tadokoro O, Heyeraas KJ, Kozawa Y, Vandevska-Radunovic V (2000) Neuroendocrine cells in Malassez epithelium and gingiva of the cat. Acta Odontol Scand 58:107–112

Peters BH, Peters JM, Kuhn C, Zoller J, Franke WW (1995) Maintenance of cell-type-specific cytoskeletal character in epithelial cells out of epithelial context: cytokeratins and other cytoskeletal proteins in the rests of Malassez of the periodontal ligament. Differentiation 59:113–126

Tachibana T (1995) The Merkel cell: recent findings and unresolved problems. Arch Histol Cytol 58:379–396

Tachibana T, Yamamoto H, Takahashi N, Kamegai T, Shibanai S, Iseki H, Nawa T (1997) Polymorphism of Merkel cells in the rodent palatine mucosa: immunohistochemical and ultrastructural studies. Arch Histol Cytol 60:379–389

Tadokoro O, Maeda T, Heyeraas KJ, Vandeyska-Radunovic V, Kozawa Y, Kvinnsland IH (2002) Merkel like cells in Malassez epithelium in the periodontal ligament of cats. J Periodontal Res 37(6):456–463

Morphological Changes in Merkel Cells of Mucosa Underlying Dentures

K. Ishizaki[1], C.M. Tazaki[2], C.T. Inoue[3], C.M. Shimono[4], C.K. Sakurai[1]

[1]Department of Complete Denture Prosthodontics, Tokyo Dental College
[2]Department of Physiology, Tokyo Dental College
[3]Department of Clinical Pathophysiology, Tokyo Dental College
[4]Department of Pathology, Tokyo Dental College

Summary

In this study, we investigated Merkel cells which are found in large numbers in the palatal rugae area functioning as mechanoreceptors. The morphological changes occurring as a consequence of the abnormal mechanical stimulation exerted by the denture base over the palatal mucosa were observed. Degeneration of Merkel cells and sensory nerve fibers was recognized on the first day after the denture plate was placed. Thus, continuous mechanical stimulation, even for a very short term can influence the nature and property of the sensory nerve fibers. Further, necrosis or apoptosis of Merkel cells was induced by strong continuous mechanical stimulation exerted by the denture base on the underlying mucosa.

Introduction

Merkel cells (MCs) have been found in large numbers in the palatal rugae (Van der Welf et al. 1982; Arvidsson et al. 1995), and along with other terminal sensory nerves they have been considered to function as mechanoreceptors of the palate (Van der Welf et al. 1982; Liem et al. 1984; Tachibana et al. 1991; Arvidsson et al. 1995).

Nowadays in dental practice, for those clinical cases of either complete or partially edentulous patients who have lost large numbers of teeth, in order to achieve retention and stability of the dentures, the majority of the palate is usually covered by the denture base. As a consequence, the physiological pressure exerted by the tongue and alimentary bolus over the palatal mucosa during the masticatory process is intercepted by the presence of the denture base. On the other hand, the denture base exerts a continuous mechanical stimulation over the palate. Thus, taking into consideration the physiological function of the palate, it is easy to presuppose that this situation might cause some undesired problems.

In previous studies we have investigated the distribution of MCs in the lateral palatal rugae of hamsters, and demonstrated that the number of MCs decreased as a consequence of mechanical stimulation exerted by the denture base over the palatal mucosa.

Nevertheless, in the previous experiments not only could we not control the pressure exerted over the palatal mucosa, but it was also not possible to avoid the chemical effects of the materials used for the confection of the experimental dentures.

Therefore, in order to clarify the morphological changes occurring in MCs as well as in the surrounding tissue as a consequence of mechanical stimulation exerted by the denture base over the palate, we improved the former experimental method.

Materials and Methods

Ten male golden hamsters, 8–9 weeks old, were used. 0.15 ml of pentobarbitone anesthesia was intraperitonealy injected and impressions of the palate were taken in order to make the casts. Subsequently, a 0.8-mm-deep hole was drilled over a cast at the left side of the palatal rugae in order to create an elevation on the internal surface of the dentures. The experimental denture bases were made of adhesive resin (4-META/MMA-TBB) over the cast, and were set on the palate of the animals under general anesthesia. The denture bases were pressed against the palatal mucosa and fixed to the posterior teeth using the same adhesive resin (Fig. 1). The animals were killed 1, 3, 5 and 7 days later.

Results and Discussion

Figure 2 shows histological images (sectioned at 1 μm thickness) of the palatal rugae area which had been subjected to stronger pressure. Histological images showed an extremely thin epithelium and a reduction of the epithelial ridges. Thus, it was demonstrated that palatal rugae in this area were strongly pressed by the elevation created on the denture base.

Figure 3a shows an electron microscope image of a normal MC containing a large number of dense core granules in the cytoplasm. Moreover, the terminal of the sensory nerve fibers making synaptic contact with the MC is shown. Figure 3b, c demonstrate changes in the MCs 3 days after the experiment. A decreased number of dense core granules in the cytoplasm can be seen in Fig. 3b, while Fig. 3c shows degeneration of the sensory nerve fibers and terminals, making synaptic contact with MCs. The same tendency was already recognizable in images taken on the first day of denture base wearing. In Figure 3d one recognizes

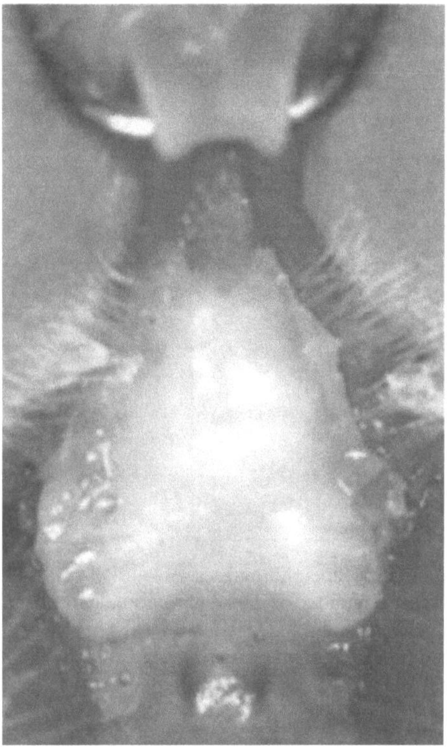

Fig. 1. The denture base fixed on the palate using adhesive resin (4-META/MMA-TBB)

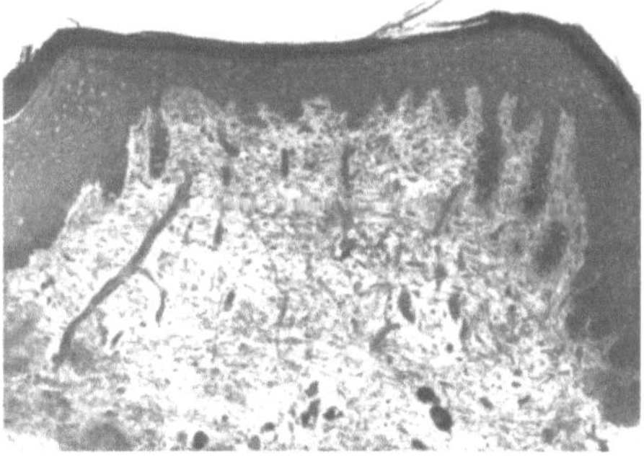

Fig. 2. Light microscopic photograph of the depressed palatal rugae. Thin epithelium and a reduction of the epithelial ridges were observed in the experimental group

damage to the cell membrane and intracellular infiltration of the surrounding tissues, leading to distortion of the cell's shape on day 5 after the experiment.

In the experimental group, a decreased number and, in some instances, a complete disappearance of dense core granules in the cytoplasm of MCs was found.

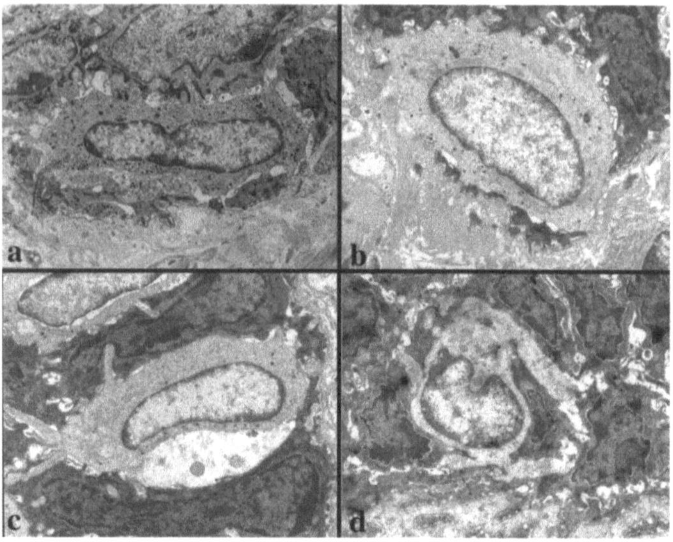

Fig. 3a–d. Electron microscope photographs of Merkel cells (*MC*). **a** Normal MC in the control group. **b** Decreased number of dense core granules in the cytoplasm was observed in MCs 3 days after the experiment. **c** Degeneration of sensory nerve terminals was seen in this specimen obtained 3 days after the experiment. **d** MCs without any granules were observed in this specimen obtained 5 days after the experiment

These findings suggest that continuous mechanical stimulation, even of a low intensity, induces morphological changes in the exposed MCs. Degeneration of sensory nerve fibers occurs on the first day of denture base wearing. Obviously, continuous mechanical stimulation, even for a very short term, can influence the nature of sensory nerve fibers. Since cells suffered damage to their membrane leading to intracellular infiltration in the surrounding tissues, continuous mechanical stimulation exerted by the denture base on the mucosa underlying the denture appears strong enough to cause a decrease in the number of MCs.

References

Arvidsson J, Fundin BT, Pealler K (1995) Innervation of the hard palate in the rat studied by anterograde transport of horseradish peroxidase conjugates. J Comp Neurol 351:489–498

Liem RSB, van Willigen JD, Jansen HWB (1984) Merkel cell neurite complexes in the hard palate of the Wistar rat (*Rattus norvegicus*). Arch Oral Biol 29:249–251

Tachibana T, Fujiwara N, Nawa T (1991) Mechanoreceptors of the hard palate of the Mongolian gerbil include special junctions between epithelia and Meissner lamellar cells: a comparison with other rodents. Anat Rec 231:396–403

Van der Welf AAD, Liem RSB, Jansen HWB (1982) The innervation pattern of the rugae on the hard palate of the Wistar albino rat. Acta Morph Neerl Scand 20:283–292

A Study on the Sensory System of Peri-Implant Tissue – Behavior of Merkel Cells and Nerve Fibers

Y. Suzuki[1], T. Inoue[2], M. Tazaki[3], M. Shimono[4], and T. Sato[1]

[1]Department of Cr and Br Prosthodontics, Tokyo Dental College, Tokyo, Japan
[2]Department of Clinical Pathophysiology, Tokyo Dental College, Tokyo, Japan
[3]Department of Physiology, Tokyo Dental College, Tokyo, Japan
[4]Department of Pathology, Tokyo Dental College, Tokyo, Japan

Summary

The purpose of the study is to investigate the behavior of Merkel cells and nerve tissue in peri-implant tissue using immunohistochemistry, electron microscopy, and confocal laser scanning microscopy. These results suggested that even though the regeneration of epithelium and nerve tissue occurred around the peri-implant region, Merkel cells did not regenerate in peri-implant. Therefore, it has to be considered that MC–neurite complexes are not formed in the peri-implant tissue.

Introduction

Previous studies revealed that Merkel cells (MCs) are widely distributed in the oral cavity such as lip, hard palate, mandibular gingiva, and buccal mucosa (Barrett et al. 2000). In the fundamental ultrastructural study of MCs in the rodent oral cavity, there are two types of MCs, dendritic and roundish in shape. Tachibana et al. (1997) reported that most dendritic MCs were not innervated, but roundish MCs were innervated (Tachibana 1995; Tachibana et al. 1997). Furthermore, MC – neurite complexes have been considered as slowly adapting type I mechanoreceptors in oral mucosa (Tazaki et al. 2000). There are contradicting reports with regard to the origin of MCs. Moll et al. (1996) reported that MCs were post-mitotic cells to be renewed from undifferentiated keratinocytes with stem cell characteristics (Moll et al. 1996). On the other hand, Grim and Halata (2000) reported that MCs were derived from neural crest cells shown in the limb primordium of chick/quail chimeras (Grimm and Halata 1996).

However, there are a few studies on the behavior of MCs in relation to wound healing (Tachibana and Ishizeki 1981). The purpose of this study is to investigate the relationship between MCs and nerve tissue regeneration after implantation using immunohistochemistry, electron microscopy and confocal laser scanning microscopy.

Materials and Methods

Twenty-one male Syrian golden hamsters, 5 weeks of age, were used in the experiment. Under the general anesthesia, the left-side epithelium of the third palatine ruga was removed with a modified injection needle (1.2 mm in diameter). Then the palatal bone was removed with a round bar, 0.8 mm in diameter cooled with phosphate buffered saline (PBS). The bone cavity for implantation was prepared by drilling with a dental reamer (#55-#60). A Ti-6Al-4 V screw implant, 1.3 mm in diameter and 8.0 mm in length was fixed into the bone cavity using a screwdriver. During the experimental periods, animals were fed with milk only. The right-side third ruga was used as control.

For light microscope observation, the animals were sacrificed 1, 2, 3, 4, 5, 6, and 7 days after the implantation. The maxilla with implant was removed and fixed in 4% paraformaldehyde for 3 days and decalcified with 10% EDTA for 10 days. The specimens were dehydrated with graded ethanol and embedded in paraffin. Paraffin sections, approximately 2 or 3 μm in thickness were cut. For immunohistochemical analysis, sections were incubated in a 1:30 dilution of monoclonal antibody against CK 20 which reacts with MC, and were incubated in a 1:50 dilution of monoclonal antibody against NSE which reacts with nerve fibers.

For electron-microscopic observation, the animals were fixed by intracardiac perfusion with Karnovsky's fixative solution. Then the tissues were decalcified with 10% EDTA for 10 days, postfixed with 2% osmium tetroxide solution (OsO_4), and then dehydrated and embedded in epoxy resin. Ultrathin sections approximately 70 or 80 nm in thickness were cut. Thin sections were stained with uranyl acetate and lead citrate and examined using an H7100 electron microscope.

For confocal laser scanning microscopic observation, the animals were fixed and decalcified using the same method as for light microscopic observations. Specimens were rapidly frozen in liquid nitrogen and cut into sections of approximately 20 μm using a cryostat. Then the specimens were washed with PBS, and incubated with either a monoclonal antibody against CK 20 diluted 1:30 for 2 h or a polyclonal rabbit antibody against protein gene product (PGP) 9.5 diluted 1:50 for 2 h. The nuclear areas of the cells were labeled by incubation in a 1:200 dilution of TO-PRO-3 iodide for 2 h.

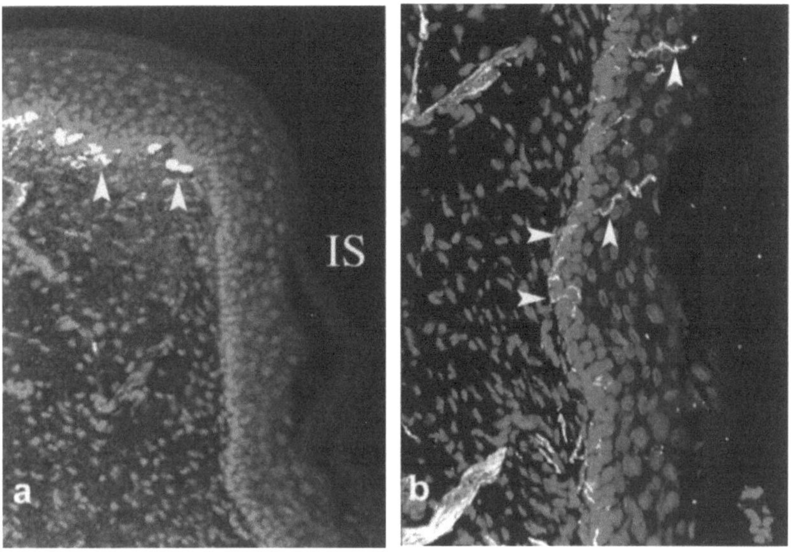

Fig. 1a, b. Seven days after the implantation. **a** *IS* Implant space, *arrowheads* MCs in the normal epithelium. **b** PGP9.5-positive nerve fibers are seen in the regenerative connective tissue and epithelium (*arrowheads*)

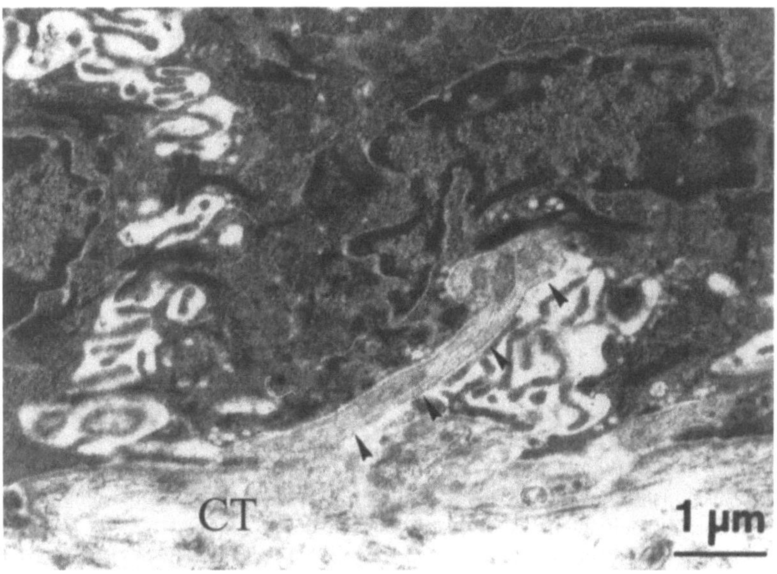

Fig. 2. Seven days after the implantation. Regenerating nerve fiber (*arrowheads*) are seen in the connective tissue just underneath the basal layer. *CT* Connective tissue

Results

In the control group, abundant CK 20-positive cells were observed in the basal layer of the epithelial rete ridge. In addition, NSE-positive cells were seen in the connective tissue just underneath the basal layer. Ultrastructurally, many MCs with plenty of neurosecretory granules were seen in the basal layer and also around the MCs, nerve endings were seen making synaptic contacts. Confocal laser scanning microscopic observation showed that PGP9.5-positive fibers were seen extending from connective tissue towards the CK 20-positive cells.

In the experimental group, no MCs in the regenerative epithelium were observed until 7 days using immunohistochemistry, electron microscopy and confocal laser scanning microscopy (Fig. 1a). NSE-positive nerve fibers appeared in the regenerative connective peri-implant tissue around 4 days after the implantation. Electron-microscopic observation showed regenerating nerve fibers just below the basal cell layer from 5 days after implantation onwards (Fig. 2). Using double-labeling confocal laser scanning microscopic observation, PGP9.5-positive nerve fibers were seen in the connective tissue at 4 days after the implantation and in the epithelium at 7 days after the implantation (Fig. 1b).

Conclusion

These results suggest that even though regeneration of epithelium and nerve tissue occurs around the peri-implant region, MCs do not reappear. Thus, it is concluded that MC-neurite complexes are not formed in the peri-implant tissue.

References

Barrett AW, Cort EM, Patel P, Berkovitz BKB (2000) An immunohistological study of cytokeratin 20 in human and mammalian oral epithelium. Arch Oral Biol 45:879–887

Grim M, Halata Z (2000) Developmental origin of avian Merkel cells. Anat Embroyl 202:401–410

Moll I, Zieger W, Schmelz M (1996) Proliferative Merkel cells were not detected in human skin. Arch Dermatol Res 288:184–187

Tachibana T (1995) The Merkel cell: resent findings and unsolved problems. Arch Histol Cytol 58:379–396

Tachibana T, Ishizeki K (1981) Merkel cell development in the wound healing in the labial mucosa of adult rabbits. Arch Histol Jpn 44:151–165

Tachibana T, Yamamoto H, Takahashi N, Kamegai T, Shibanai S, Iseki H, Nawa T (1997) Polymorphism of Merkel Cells in the Rodent Palatine Mucosa: immunohistochemical and ultrastructural studies. Arch Histol Cytol 60:379–389

Tazaki M, Tazaki Y, Inoue T, Shimono M (2000) Polymorphism of Merkel cells from slowly adapting type I mechanoreceptors in the buccal pouch of the golden hamster. Biomedical Res 21:45–48

Merkel Nerve Endings in Sinus Hairs of Young and Aged Rats

Zdenek Halata and Klaus Sames

Institute of Anatomy, University of Hamburg, UKE Martinistr. 52, 20246 Hamburg, Germany

Summary

Whiskers are highly sensitive touch organs in all mammals apart from man. A blood sinus encloses the hair follicles completely or almost completely. One sinus hair has several thousand sensory nerve endings. Most of them are located in the epithelial thickening below the sebaceous gland. A large number of Merkel cells with associated nerve endings can be found in that region. The absolute number depends on the species and size of the hair. In cat and rat, large whiskers can have up to 2000 Merkel cell-nerve endings innervated by approximately 100 myelinated axons.

Within the epithelial thickening, Merkel cells are located in the basal epithelial layer of the follicle and arranged like scales in a pine cone. The discoid nerve terminals are located on that side of the Merkel cell opposite to the basal membrane. Synaptic contacts between Merkel cells and nerve terminals can be seen and the cytoplasm towards the nerve terminals contains numerous osmiophilic granules.

In the literature there are studies on structure, development, de- and regeneration of sinus hairs; however, we are not aware of any systematic study on age-related changes.

This study was carried out on rats aged between 2 days and 30 months, which is a high age in rats. All types of nerve endings were found both in young and old rats. The number of Merkel nerve endings depends on the size of the hair follicles. In young animals (2 days), they are still growing and mitoses are often seen in the thickening of the hair bulb (Merkel cell cuff). At this age, not all Merkel cells already have contact with nerve terminals. The number of dense core granules in Merkel cells increases during innervation. At the age of 19 days and later, almost all Merkel cells are innervated. Absolute numbers of Merkel nerve endings cannot be determined using electron microscopy. However, their ultrastructural appearance did not show any significant change with age.

Introduction

Sinus hair are also described as whiskers, vibrissae, touch hairs or contour hairs (Halata 1975). They were first described by Merkel in 1880. The hair follicles including hair bulbs are almost completely embedded in a blood sinus encased in a dense connective tissue capsule. The lower third of the blood sinus has septae of connective tissue and is called the cavernous sinus, while the upper part of the sinus extending to the sebaceous gland is one large cavity called the ring sinus. These structures separate a large number of different mechanoreceptors from the surrounding dermis and may act as a buffer against mechanical disturbance from the surrounding area. Small hair follicles sometimes only have the ring sinus and are referred to as semi-sinus hairs (Munger and Halata 1983).

Sinus hairs can be found in all mammals excluding man. Apart from a few exceptions, they are all innervated by the trigeminal nerve and positioned around the face openings, especially on the upper lip. Striated muscle fibres of the mimic muscles insert on the capsule of the blood sinus and enable the animals to move the hairs under voluntary control.

The sinus hairs are important for orientation of the animal and contain several types of mechanoreceptors: lanceolate nerve endings, small lamellated corpuscles and free nerve endings in the connective tissue between blood sinus and hair follicle epithelium and Merkel nerve endings in the outer root sheet of the hair follicle. Merkel nerve endings are by far the most abundant of these mechanoreceptors (Halata and Munger 1980). They can be found in two different positions: superficially near the rete-ridge collar close to the penetration of the hair shaft through the epidermis and in the epithelial thickening of the hair follicle below the sebaceous gland. In this location, the basal lamina is relatively thick and visible under light microscopy. This has led to the name glassy membrane (Halata and Munger 1980).

In the rete ridges, the Merkel nerve endings have the same structure as in touch domes (Iggo and Muir 1969; Halata 1975) or in the rete pegs of glabrous skin (Halata 1971). Merkel cells in the thickening of the external root sheath of the hair follicle are oval or angular in shape and oriented with a long axis at an acute angle to the longitudinal axis of the hair shaft. The nerve endings associated with these Merkel cells are found on the internal root sheath aspect of Merkel cells, thus polarized away from the connective tissue compartment of the sinus hair. Between Merkel cells and nerve terminals synaptic contacts can be seen. The surface of the Merkel cells is enlarged by finger-like cytoplasmic protrusions extending between neighbouring keratinocytes or through the glassy membrane into the blood sinus. There are various reports in the literature describing structure, development and function of sinus hairs in various mammalian species (for review, see Zelena 1994). However, we are not aware of any systematic study on age-related changes of sinus hairs and their innervation. Thus, this is the first report examining Merkel nerve endings in rats of different age groups.

Materials and Methods

Rats aged 2 days, 19 days and 30 months were investigated. In each age group four animals were killed with an overdose of pentobarbital, the chest was opened and glutardialdehyde infused into the left ventricle. Sinus hairs of the upper lip were excised and left in the same fixation solution for between several hours and 2 days. Before postfixation in 1% OsO_4 for 1 h, the tissue was thoroughly washed in phosphate buffer solution. Following dehydration the tissue was embedded in Epon 812. Four hair follicles were taken from each animal for semithin sections and stained according to Laczko and Levai (1975). Selected blocks were trimmed for ultrathin sectioning, contrasted according to Reynolds (1963) and examined under the electron microscope (Philips 300).

Results

2-Day-Old Rats

The sinus hairs are small (Fig. 1a) and completely surrounded by a blood sinus which is not yet divided into ring sinus and cavernous sinus. The sebaceous gland is small and not fully developed. Below is a thickening of the hair follicle containing Merkel cells and nerve terminals (Fig. 1b). The basal lamina is thinner than in fully developed hair follicles.

Merkel cells are oval or polygonal in shape with their long axis oriented perpendicular to the hair shaft. In the upper and medial parts most Merkel cells make close contact with discoid nerve terminals, while some Merkel cells have no nerve terminal in their vicinity. In some sections, branching of axons can be seen supplying several neighboring Merkel cells (Fig. 1b). The osmiophilic granules with diameters of 600–800 nm could be found in all parts of the cytoplasm with a slight concentration in the part facing the nerve terminal. Not all afferent axons are myelinated. In the thickening of the hair follicle epithelium in close vicinity of the Merkel cells, mitoses are often observed (Fig. 2a).

19-Day-Old Rats

The sinus hairs are larger with clearly visible separation into a ring sinus and cavernous sinus. The sebaceous gland is fully developed with more pronounced thickening below. The basal lamina is about twice as thick as in the 2-day-old rats.

The Merkel cells are similar in structure to those in the 2-day-old animals (Fig. 2b). All Merkel cells seen by us have contact with nerve terminals. The number of granules has increased and they are clearly concentrated towards the nerve terminal. Moreover, the discoid nerve terminals are thicker and contain more mitochondria. Clear vesicles with diameters of 400–600 nm can be seen in the axoplasm, particularly in the part adjacent to the Merkel cell. The afferent axons are completely myelinated with diameters around 6 μm.

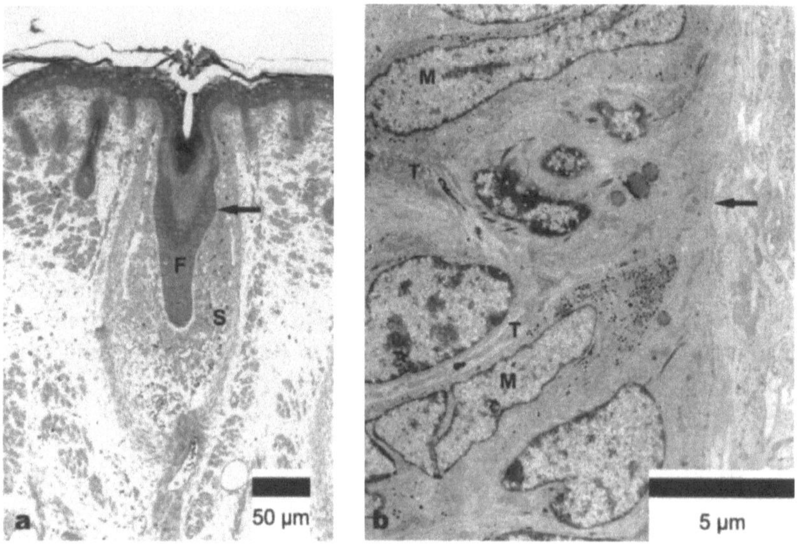

Fig. 1. a Sinus hair of a 2-day-old rat in longitudinal semithin section. The hair follicle (*F*) is embedded in a blood sinus (*S*) not yet separated into cavernous and ring sinus. The bulge region is slightly thicker (*arrow*)

b Sinus hair of a 2-day-old rat in longitudinal ultrathin section. Two Merkel cells (*M*) innervated by branched nerve terminals (*T*) from the bulge region. The dense core granules are not yet oriented towards the nerve terminal. The basal lamina (*arrow*) is relatively thin

30-Month-Old Rats

The hair follicles are large and up to 1 mm thick and several mm long. The blood sinus consists of a large ring sinus in the upper two thirds and a smaller cavernous sinus in the lower third mainly surrounding the hair bulb. The thickening below the sebaceous gland is well developed forming a conical shape towards the surface. The epithelium of the basal layer in the outer root sheath is clearly different from the remaining epithelium: the cytoplasm contains only sparse tonofilaments, the nuclei are oval with their long axes perpendicular to the hair shaft (Fig. 3a). The basal lamina (glassy membrane) is about 8 µm thick. The Merkel nerve endings are located in the basal layer of the outer root sheath. There are always pairs of Merkel cells and nerve terminals (Fig. 3b). The number of mitochondria within the nerve terminals is similar to that of the 19-day-old rats. The Merkel cells are oval or triangular with a lobulated nucleus and osmiophilic granules, especially in that part of the cytoplasm facing the nerve terminal. Swollen mitochondria are often seen which might be the result of slower diffusion of the fixative in these larger hair follicles.

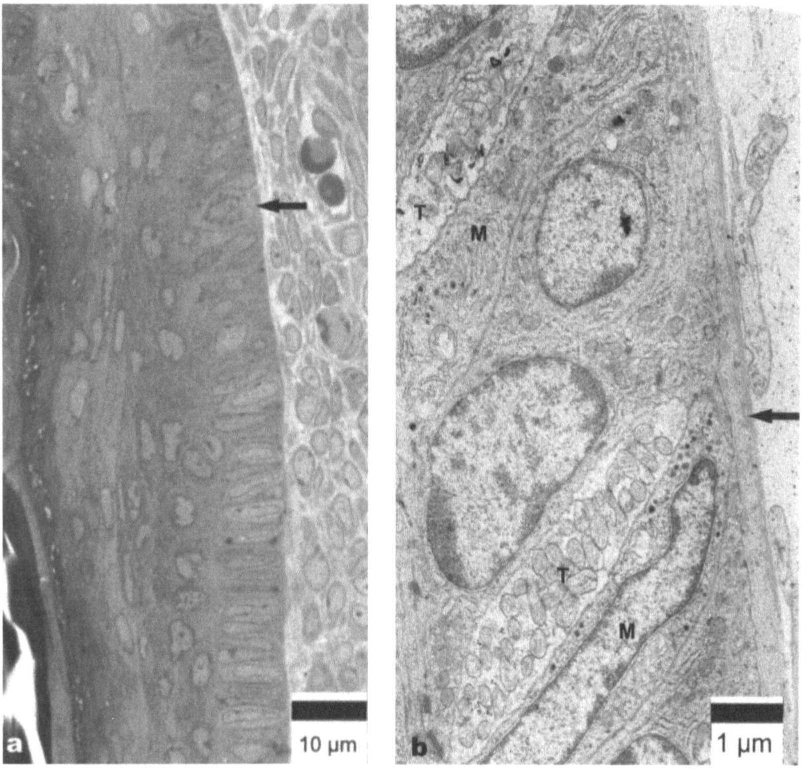

Fig. 2. **a** Sinus hair of a 2-day-old rat in longitudinal semithin section. The bulge region contains a row of Merkel cells. A mitosis within this row is indicated by *arrow*. **b** Sinus hair of a 19-day-old rat in longitudinal ultrathin section. Two Merkel cells (*M*) with associated nerve terminals (*T*) from the bulge region are shown. The dense core granules in the cytoplasm of the Merkel cells are mostly close to the nerve terminals. The basal lamina (*arrow*) is thicker than in **b**

Discussion

Most developmental studies are concerned about the central representation of sinus hairs in the somatosensory cortex (barrels; Woolsey and Van der Loos 1970; Munger and Rice 1986; Rice and Munger 1986; Rice et al. 1986) or with reinnervation after peripheral denervation (Hartschuh and Weihe 1977a, b). Further details are reviewed by Zelena (1994). We are not aware of any systematic study of the structure of sinus hairs during ageing. The present study demonstrates clear differences particularly in the size of the hair follicle and the blood sinus; both increase with age. This goes together with an increase in the number of Merkel nerve ending in the outer root sheath. While in rat and small mammals the number of Merkel cells is up to 2,000, it has been shown that in large aquatic mammals

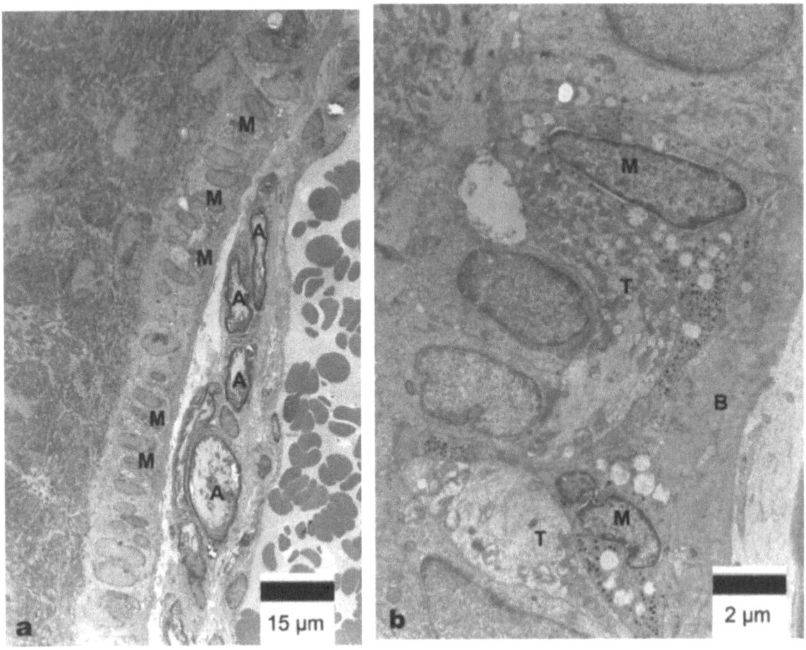

Fig. 3. a Sinus hair of a 30-month-old rat in longitudinal ultrathin section. Bulge region with several Merkel cells (*M*). Myelinated axons are marked with *A*. **b.** Sinus hair of a 30-month-old rat in longitudinal ultrathin section. Higher magnification than in **a**. Two Merkel cells (*M*) are innervated by branched nerve terminals (*T*). The basal lamina (*B*) is thick (glassy membrane)

one sinus hair can have up to 20,000 Merkel nerve endings (Halata 1975; see Dehnhardt et al., this Vol.).

The increase in the number of Merkel cells is reflected in the finding of mitoses in the basal layer of the epithelium below the sebaceous gland in 2-day-old rats – most likely of Merkel cell precursors. Such mitoses were not seen in older animals of 19 days or 30 months. This is in line with the findings of Munger and Rice (1986) that the development of Merkel cells and associated nerve endings is not complete at birth, but extends into the postnatal age.

Similar to other species, one axon supplies several Merkel nerve endings (Van Horn 1970; Stephens et al. 1973; Halata and Munger 1980). According to Ebara et al. (see Ebara et al., this Vol.), one axon supplies up to 200 Merkel cells. These are arranged in columns in the rat while in the cat they form patches. Innervation occurs prenataly and carries on postnataly. Myelination is not completed until some time after birth (Zelena 1994). In the present study, not all Merkel cells of 2-day-old rats were innervated. In contrast, in 19-day and 30-month-old rats, all Merkel cells had contact with nerve terminals.

During electron microscopy, there was an obvious increase in the number of granules between noninnervated Merkel cells and those in contact with nerve terminals. While the granules were relatively evenly distributed throughout the cytoplasm, there was a clear accumulation of the dense core granules towards the nerve terminal after the contact had been established. The shape of Merkel cells changed slightly from oval in young rats to flattened in older rats. Otherwise the ultrastructure of Merkel cells did not show significant changes with age.

Acknowledgements

The authors would like to thank Prof. Niedermüller, Institute of Physiology, University of Veterinary Medicine, and Ludwig Boltzmann Institute for Aging Research – Biogerontology, Vienna, Austria, for generously providing the samples of the old rats. Excellent technical assistance by Ms. Brigitte Asmus is appreciated.

References

Halata Z (1971) Innervation der unbehaarten Nasenhaut des Maulwurfs (*Talpa europaea*). I. Intraepidermale Nervenendigungen. Z Zellforsch 125:108–120

Halata Z (1975) The mechanoreceptors of the mammalian skin. Ultrastructure and morphological classification. Adv Anat Embryol Cell Biol 50:1–77

Halata Z, Munger BL (1980) Sensory nerve endings in rhesus monkey sinus hairs. J Comp Neurol 192:645–663

Hartschuh W, Weihe E (1977a) The effect of denervation on Merkel cells in cats. Neurosci Lett 5:327–332

Hartschuh W, Weihe E (1977b) The effect of denervation on Merkel cells in nude mice and cats. J Cut Pathol 4:221

Iggo A, Muir AR (1969) The structure and function of a slowly adapting touch corpuscle in hairy skin. J Physiol 200:763–796

Laczko J, Levai G (1975) A simple differential staining method for semi-thin sections of ossifying cartilage and bone tissue embedded in epoxy resin. Mikroskopie 31:1–4

Merkel F (1880) Ueber die Endigungen der sensiblen Nerven in der Haut der Wirbeltiere. Schmidt, Rostock

Munger BL, Halata Z (1983) The sensory innervation of primate facial skin. I. Hairy skin. Brain Res Rev 5:45–80

Munger BL, Rice FL (1986) Successive waves of differentiation of cutaneous afferents in rat mystacial skin. J Comp Neurol 252:404–414

Reynolds ES (1963) The use of lead citrate at high pH as an electron-opaque staining in electron microscopy. J Cell Biol 17:208–212

Rice FL, Munger BL (1986) A comparative light microscopic analysis of the sensory innervation of the mystacial pad. II. The common fur between the vibrissae. J Comp Neurol 252:186–205

Rice FL, Mance A, Munger BL (1986) A comparative light microscopic analysis of the sensory innervation of the mystacial pad. I. Innervation of vibrissal follicle-sinus complexes. J Comp Neurol 252:154–174

Stephens RJ, Beebe JJ, Poulter TC (1973) Innervation of the vibrissae of the California sea lion, *Zalophus californianus*. Anat Rec 176:421–442

Van Horn RN (1970) Vibrissae structure in the rhesus monkey. Folia Primat 13:241–285

Woolsey TA, Van der Loos H (1970) The structural organisation of layer IV in the somatosensory region (SI) of mouse cerebral cortex. Brain Res 17:205–242

Zelena J (1994) Nerves and mechanoreceptors. Chapmann and Hall, London

Immunohistochemical Characterization of Normal Canine Merkel cells

G.A. Ramírez; P. Herráez; M. Andrada; A. Fernández; P. Castro and A. Espinosa de los Monteros

Department of Comparative Pathology, Veterinary Faculty, University of Las Palmas de Gran Canaria, Las Palmas de Gran Canaria, Spain

Summary

Cutaneous Merkel cells (MCs) have been well documented in humans, but less in other mammals. In dogs, there are only a few references about immunohistochemical characterization of MCs. We present the immunohistochemical profile of MCs in the dog for the most reliable antibodies used in human medicine. Tissue samples from several locations were obtained from five adult dogs of both sexes and different age and breed, fixed in 10% buffered formalin, embedded in paraffin wax and cut in 3-µm tissue sections. The ABC method was used with different poly- and monoclonal antibodies. Positive immunoreaction was found in MCs of hair follicles, skin and mucosae of several locations in each dog studied for anti-cytokeratins 8, 18 and 20, anti-neurofilaments, anti-chromogranin A, anti-neuron-specific enolase, and anti-synaptophysin. Immunoreaction was always cytoplasmic with differences in intensity, pattern of intracytoplasmic distribution and reaction type. Other cytokeratins, anti-S100 protein, anti-vimentin and anti-glial fibrillary acid protein were absent in canine normal MCs.

Introduction

Merkel cells (MCs) are distributed widely in the skin, hair follicles and oral mucosae of every vertebrate class (Toyoshima et al. 2000). They are localized at the basal layer of the epidermis in skin and of the epithelium in mucosae, showing a close relation with nerve endings forming the Merkel cell-neurite complex. Several functions have been proposed for this cellular population, such as slowly adapting type I mechanoreceptors (particularly important for the slowly adapting response to maintained mechanical stimuli), endocrine cells, stem cells for gustatory cells in taste buds, or cells inducing outgrowth of nerve fibres (Ogawa 2000).

Merkel cells were first described in 1875 by Friedrich Siegmund Merkel in mammalian and avian skin using a staining method based in osmium tetroxide (Merkel 1875). Nowadays, MCs have been demonstrated in many species by different procedures, mainly immunohistochemistry and electron microscopy methods. Different antibodies have been used for immunohistochemical staining of MCs. In dogs, they have been poorly studied and only few reports of their immunophenotype have been reported (Hartschuh et al. 1983, 1984; Cheng Chew and Leung 1992, 1993).

The aim of this study was to determine the immunohistochemical profile of Merkel cells in the dog, using the most used commercial antibodies reported in the literature to characterize these cells.

Materials and Methods

Tissue specimens were obtained from five dogs of different breeds (boxer, mixed-breed dog, two poodles and cocker spaniel), ages (between 2 and 4 years) and both sexes (two males and three females). Samples were taken from several skin locations and from oral, nasal, ocular, anal and genital mucosae. They were fixed in 10% buffered formalin for 24 h, embedded in paraffin and cut into 3-μm sections. The different sections were deparaffined, rehydrated in graded alcohols, and incubated with 3% hydrogen peroxidase in methanol for 30 min to block endogenous peroxidase activity. Antigen retrieval treatments and times are shown in Table 1. The slides were covered with 10% normal goat serum in PBS for monoclonal antibodies and in 10% normal swine serum for polyclonal antibodies for 30 min before incubation with the primary antibodies (Table 1) for 18 h at 4 °C. The avidin-biotin peroxidase complex (ABC) method was applied (Vector Laboratories, Burlingame, CA) and the chromogen used was 3-amino-9-ethylcarbazole. The slides were counterstained with Mayer's hematoxylin.

The immunostainings were evaluated separately and results were based upon the consensus of at least two observers. The intensity of the immunoreaction was evaluated on a three-point scale (1*i*: weak, 2*i*: moderate, 3*i*: intense).

Results

Different patterns of immunoreactivity and reaction intensity observed are shown in Table 1. The immunoreaction was always located in the cellular cytoplasm, although every antibody presented characteristic proprieties.

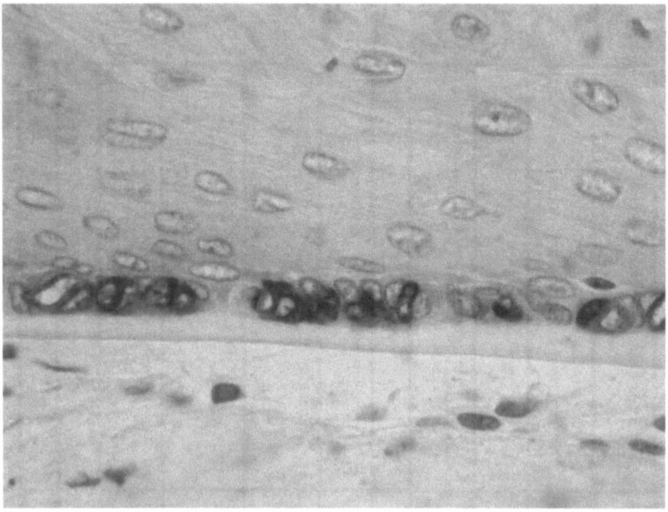

Fig. 1. Dog. Vibrissae. Cytokeratin 20 antibody staining reaction homogeneously distributed within the cytoplasm of a group of Merkel cells. ABC method; AEC chromogen; Mayer's hematoxylin counterstaining, ×1000

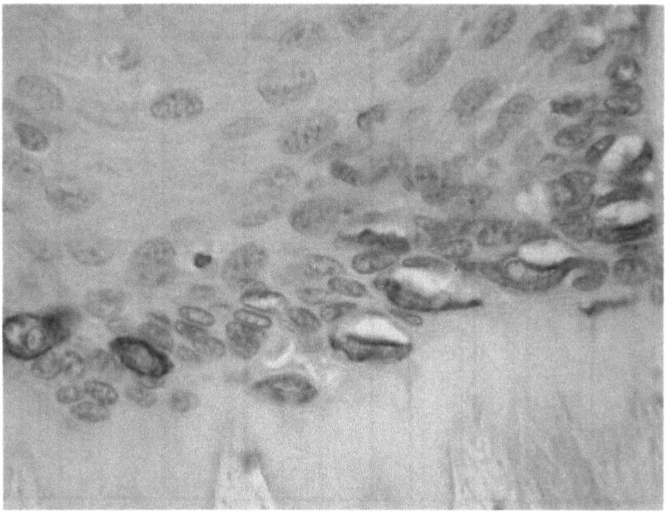

Fig. 2. Dog. Tylotrich hair follicle. Neurofilaments antibody staining reaction homogeneously distributed within the cytoplasm of Merkel cells. ABC method; AEC chromogen; Mayer's hematoxylin counterstaining, ×1000

Table 1: Antibodies procedures and results of the immunohistochemical study of canine Merkel cells.

Antibodies	Type	Specificity	Epitope retrieval	Time (min.)	Dilution	Reaction intensity	Immunoreactive patterns
Ks 20.8*	Mab	Cytokeratin 20†	Citrate pH 6 100°C	6	1:50	3i	homogenous/granular and cytoplasmic (membrane reinforcement occasionally)
Keratin**	Pab	Human skin cells cyto-keratins	-	-	1:150	-	-
NCL 5D3**	Mab	Cytokeratins 8 and 18†	Citrate pH 6 100°C	6	1:20	3i	Homogeneous and cytoplasmic
RCK 102**	Mab	Cytokeratins 5 and 8†	Protease	10	1:20	-	-
NSE*	Mab	Human gg-enolase	Citrate pH 6 100°C	5	1:1000	3i	granular and cytoplasmic
Chromogranin A*	Pab	Human chromogranin A	Citrate pH 6 100°C	6	1:100	3i	Granular/homogeneous and cytoplasmic (neurite cellular side)
Synaptophysin*	Mab	Human synaptophysin	Citrate pH 6 100°C	6	1:25	2i	Granular and cytoplasmic (opposite to neurite cellular side)
Neurofila-ment**	Mab	70 and 200 kD neuro-filaments protein	Citrate pH 6 100°C	6	1:20	1i / 3i	Homogeneous and cytoplasmic
GFAP**	Pab	Glial fibrillary acid protein	-	-	1:150	-	-
Vimentin**	Pab	Vimentin	-	-	1:100	-	-
S-100*	Pab	Cow S100 protein	-	-	1:500	-	-

*: Dako; **: EuroDiagnostica; Mab: Monoclonal antibody; Pab: Polyclonal antibody; †: The Moll catalogue of cytokeratins. Reaction intensity: 1*i*: weak; 2*i*: moderate; 3*i*: intense.

All skin, follicular and mucosal MCs reacted with the anti-cytokeratin (CK) 20 monoclonal antibody. The immunoreaction was distributed homogeneously through the cell cytoplasm (Fig. 1), although a granular positive reaction and membrane reinforcement were occasionally seen. Intensity of the immunostaining was 3*i* in the vast majority of the slides evaluated. With anti-CKs 8 and 18 monoclonal antibody, the immunoreaction was positive for all MC types as well, always showing an homogeneous cytoplasmic pattern with a 3*i*-reaction intensity.

The anti-NSE monoclonal antibody showed an intense granular positive reaction (3*i*) in all MC types distributed in the whole cytoplasm. There was a difference in intensity for the anti-NF monoclonal antibody. Whereas follicular MCs showed an intense homogeneous cytoplasmatic positive reaction (3*i*; Fig. 2), skin and mucosal MCs appeared with a weak reaction (1*i*). The immunoreaction patterns observed with both anti-CgA polyclonal antibody and anti-synaptophysin monoclonal antibody were interesting. Whereas with the former the reaction was located in the basal part of the cytoplasm next to the neurite, with the latter the reaction was present in the opposite cellular portion. The intensities of immunostaining were 3*i* for CgA and 2*i* for synaptophysin. Occasionally, CgA reaction was distributed in the entire cytoplasm.

Anti-vimentin, anti-keratins, anti-GFAP and anti-S-100 protein polyclonal antibodies and anti-CKs 5 and 8 monoclonal antibodies did not show any reaction in MCs.

Discussion

The immunohistochemical profile of MCs has been widely studied in humans and in some animals. Many reports about immunoreactions against intermediate filaments, peptides and other substances have been described. In dogs, we have found only a few references of immunohistochemical demonstrations of vasoactive intestinal polypeptide (VIP; Hartschuh et al. 1983, 1984), calcitonin gene-related peptide (CGRP; Cheng Chew and Leung 1993) and met- and leu-enkephalin (Cheng Chew and Leung 1992).

Cytokeratin 20 (CK20) is a specific marker of cutaneous Merkel cells (Moll et al. 1995). It has been successfully demonstrated in mouse skin (Moll et al. 1996) and in rodent oral mucosae (Tachibana et al. 1997). We found CK20 in skin, mucosae and hair follicular MCs in the dog showing more similar characteristics than previous reports. An intense immunostaining was also obtained with anti-CKs 8 and 18 antibodies at the same locations. Neuron-specific enolase (NSE) has been reported to be present in MCs of nose of cats and rats in a low proportion (1 out of 20 or 30; Gu et al. 1981). In our study, NSE was demonstrated in almost all locations with a very intense immunoreaction. Human MCs also contain neurofilaments (NF; Narisawa et al. 1994), although they have not been found in lip skin of rabbits (Saurat et al. 1984) and other mammals. In dogs, NF was widely present in MCs of vibrissae and the immunoreaction was diffuse within the entire cytoplasm of the cells. The immunoreaction in haired skin and mucosal was slighter.

The immunoreaction against chromogranin A and synaptophysin showed a very particular pattern in the cell cytoplasm. While the immunoreaction obtained with anti-chromogranin A antibody was located in the basal part of the cytoplasm next to the neurite, the synaptophysin immunostaining was observed in the opposite cellular portion. These patterns accord with the ultrastructural distribution of electron-dense granules and small clear-content vesicles, respectively (García Caballero et al. 1989; Hartschuh et al. 1989) This fact suggests that secretory electron-dense core granules and small clear vesicles could be involved in two different secretory pathways of canine MCs, as has been suggested in MCs of pig-snout skin (García Caballero et al. 1989) and has been demonstrated in neurons (Lundberg and Hökfelt 1983). These results demonstrate that commercial monoclonal and polyclonal antibodies used in humans and other mammals can also be employed in the immunocharacterization of Merkel cells in the dog, and the reactions observed with these antibodies are similar to those previously described in other species.

References

Cheng Chew SB, Leung PY (1992) Species variability in the expression of met- and leu-enkephalin-like immunoreactivity in mammalian Merkel cell dense-core granules. A light- and electron-microscopic immunohistochemical study. Cell Tissue Res 269:347–351

Cheng Chew SB, Leung PY (1993) CGRP-immunoreactivity in Merkel cells and nonmyelinated nerve plexuses of dog skin. Neuroreport 4:457–459

García Caballero T, Cuevas J, Gallego R, Rosón E, Fortaleza J, Beiras A (1989) Synaptophysin-like immunoreactivity in the Merkel cells of pig-snout skin. Ultrastr Pathol 13:55–61

Gu J, Polack JM, Tapia FJ, Marangos PJ, Pearse AG (1981) Neuron-specific enolase in Merkel cells of mammalian skin: the use of specific antibody as a simple and reliable histologic marker. Am J Pathol 104:63-68

Hartschuh W, Weihe E, Yanaihara N, Reinecke M (1983) Immunohistochemical localization of vasoactive intestinal polypeptide (VIP) in Merkel cells of various mammals: evidence for a neuromodulator function of the Merkel cell. J Invest Dermatol 81:361-364

Hartschuh W, Reinecke M, Weihe E, Yanaihara N (1984) VIP-immunoreactivity in the skin of various mammals: immunohistochemical, radioimmunological and experimental evidence for a dual localization in cutaneous nerves and Merkel cells. Peptides 5:239–245

Hartschuh W, Weihe E, Egner U (1989) Chromogranin A in the mammalian Merkel cell. J Invest Dermatol 93:641–648

Lundberg JM, Hökfelt T (1983) Coexistence of peptides and classical neurotransmitters. Trends Neurosci 6:325–333

Merkel F (1875) Tastzellen und Tastkörperchen bei den Hausthieren und beim Menschen. Arch Mikr Anat 11:636–652

Moll I, Kuhn C, Moll R (1995) Cytokeratin 20 is a general marker of cutaneous Merkel cells while certain neuronal proteins are absent. J Invest Dermatol 104:910–915

Moll I, Paus R, Moll R (1996) Merkel cells in mouse skin: intermediate filament pattern, localization, and hair cycle-dependent density. J Invest Dermatol 106:281–286

Narisawa Y, Hashimoto K, Kohda H (1994) Immunohistochemical demonstration of the expression of neurofilament proteins in Merkel cells. Acta Dermato-Venereol 74:441–443

Ogawa H (2000) Merkel cells: possible mechanoreceptors. In: Suzuki H, Ono T (eds) Merkel cells, Merkel cell carcinoma and neurobiology of the skin. Elsevier, Amsterdam, pp 83–87

Saurat JH, Mérot Y, Didierjean L, Dahl D (1984) Normal rabbit Merkel cells do not express neurofilament proteins. J Invest Dermatol 82:641–642

Tachibana T, Yamamoto H, Takahashi N, Kamegai T, Shibanai S, Iseki H, Nawa T (1997) Polymorphism of Merkel cells in the rodent palatine mucosa: immunohistochemical and ultrastructural studies. Arch Histol Cytol 60:379–389

Toyoshima K, Seta Y, Harada H, Toyono T (2000) Morphology and distribution of Merkel cells in some vertebrates. In: Suzuki H, Ono T (eds) Merkel cells, Merkel cell carcinoma and neurobiology of the skin. Elsevier, Amsterdam, pp 83–87

Merkel Cells in the Human Fetal and Adult Esophagus

T. Schulz[1], E. Michou[1], W. Hartschuh[2], and P.H. Wuensch[1]

[1]Institute of Pathology, Klinikum Nuremberg, Nuremberg, Germany
[2]Department of Dermatology, University of Heidelberg, Heidelberg, Germany

Summary

Merkel cells in the skin and the nearby mucous membranes have been well studied. However, there are still very few data concerning neuroendocrine and Merkel cells in the esophagus. In the present study, we extended these investigations to a larger group of post-mortem whole esophagus specimens. Twenty-five specimens of adult and 25 specimens of fetal esophagus were studied using immunohistochemistry. For evaluation of the immunoprofile and innervation of the Merkel cells, various antibodies were used (against cytokeratins 18, 19, 20, chromogranin A, NSE, synaptophysin, serotonin, neurofilament). For evaluating the distribution of the Merkel cells, antibodies against cytokeratin 20 were decisive. Merkel cells could be detected in 80% of the adult and 68% of the fetal specimens. The number of Merkel cells varied markedly from case to case, reflecting a great interindividual variation. On average, there were 2.2 Merkel cells/cm esophageal mucosa in the adult and 1.2 Merkel cells/cm in the fetal specimen. The adult esophageal Merkel cells occurred as single cells located in the lower parts of the epithelial layer. Most of these Merkel cells did not possess long dendrites. No stromal Merkel cells were detected. Fetal Merkel cells first occurred in the 13th week of gestation. They were often arranged in small clusters, bearing long dendrites and being also located in the upper and uppermost layers of the epithelium. A few stromal fetal Merkel cells were apparent first occurring in the 20th week. The immunoprofile of the Merkel cells was comparable to that of Merkel cells in the human skin. No innervation of the Merkel cells could be detected. Our study supports the fact that Merkel cells exist in the human esophagus. For the first time we were able to document fetal esophageal Merkel cells. A marked interindividual variation in the distribution, and moreover, the sole presence of Merkel cells were striking, provoking some speculation concerning the function of human esophageal Merkel cells, especially as no innervated Merkel cells were found. Further studies on various animals may be helpful to answer the question of whether human esophageal Merkel cells perhaps are a phylogenetic rudiment.

Introduction

Merkel cells have been detected in the skin of many species (Hartschuh et al. 1986; Tachibana 1995) and also in the various mucous membranes near to the skin, which altogether have a multi-layered noncornifying epithelium (Fetisoff et al. 1990; Tachibana 1995; Hilliges et al. 1996). Moreover, the esophagus bears a multi-layered noncornifying epithelium. However, up to now, there are only two reports which describe neuroendocrine cells and Merkel cells in the unaltered squamous epithelium of the esophagus (Tateishi et al. 1974; Harmse et al. 1999). In our present study, we tried to resolve some questions concerning esophageal Merkel cells still not worked on in the above-mentioned studies. First, we chose a greater number of esophageal specimens, especially of fetal ones. This was because Merkel cells have not been found in fetal esophagi so far (Harmse et al. 1999; Botta et al. 2001). However, they are very numerous in human fetal skin, and therefore it would be very difficult to accept the neuroendocrine cells of the human esophagus as Merkel cells if they did not occur in fetal life. Second, we looked for innervation of the esophageal Merkel cells. Third, we applied several antibodies known to react with Merkel cells, but hitherto untested in esophageal Merkel cells. Fourth, we counted the Merkel cells per length of esophageal epithelium in order to enable a comparison with the known Merkel cell density in the skin.

Materials and Methods

Twenty-five postmortem adult esophageal whole specimens (mean age 66 years) and 25 fetal esophageal specimens from various weeks of gestation were investigated. We only took specimens free of postmortem autolytic changes. The adult specimens were divided into 15 probes from top to bottom.

The following monoclonal antibodies were used: monoclonal antibody (mAb) DC 10 (Dako) against human cytokeratin (CK)18, mAb RCK 108 (Dako) against CK 19, mAb Ks 20.8 (Dako) against CK 20, LK2 H10 (Linaris) directed against human chromogranin A, DP 12.10, DP 43.16, DP 5.1.12 (Linaris) directed against human neurofilaments of 210, 160 and 70 kDa, mAb MiG-N3 (Linaris) directed against neuron-specific enolase, mAb Z 0311 (Dako) directed against S 100, mAb 5HT-H209 (Dako) directed against serotonin, mAb Snp88 (BioGenex) directed against synaptophysin.

We performed immunohistochemistry using the streptavidin-biotin-peroxidase complex technique (Hsu et al. 1981). Antibodies against CK 20 were decisive to us for the identification of the Merkel cells (Moll et al. 1995). The adjacent cardiac mucosa of the stomach and also esophageal glands served as internal controls of our immunoreactions, as both structures express CK 20 (Moll et al. 1995; Schulz, unpubl. data). Double immunohistochemistry (Hsu and Soban 1982) was used to study innervation of the Merkel cells, using anti-CK 20 for labeling the Merkel cells and anti-neurofilament or anti-synaptophysin to visualize nerves.

Morphometrical analysis: Merkel cells were counted per length of epithelium, using digitized morphometry. Only CK20-positive cells containing the nucleus were counted, whereas cytoplasmic processes were not evaluated. The measurements were analysed using the Wilcoxon signed rank test for dependent data.

Results

Adult Specimens

We could detect evenly distributed intraepithelial Merkel cells in all parts of the esophagi. There was no statistically significant difference in the Merkel cell density from hypopharynx to cardia. However, there was a great interindividual variation in the density of Merkel cells. Twenty percent of the esophagus specimens did not contain any Merkel cells, whereas other specimens showed a considerable amount of Merkel cells. The average density of Merkel cells in the esophageal epithelium was 2.2 per cm, which is comparable to that of human adult epidermis regions without special touch properties (Moll et al. 1990).

The Merkel cells were mainly located as single cells lying in basal portions at the tips of the epithelial relief (Fig. 1). Only very sporadically did clustering of Merkel cells occur. The Merkel cells mostly were of a roundish shape (Fig. 2). No subepithelial Merkel cells could be detected. Innervation of the Merkel cells also could not be detected in double immunohistochemistry.

Concerning the immunoprofile of the Merkel cells, the best reactions, like in the skin, occurred with antibodies against CK 20. The other anti-CKs also showed positive reactions, albeit markedly weaker (not shown). Further positive reactions were obtained with anti-synaptopysin and anti-chromogranin A (not shown). The other antibodies tested remained negative.

Fetal Specimens

In contrast to the two above-mentioned reports (Harms et al. 1999; Botta et al. 2001), we were able to detect Merkel cells in fetal esophagus. Moreover, a marked interindividual variation in the number of Merkel cells was observable. Thirty-two percent of the fetal esophagi were totally devoid of Merkel cells. In general, the average density of Merkel cells was 1.2 Merkel cells per cm epithelium and therefore, lower than in the adult specimen. Merkel cells were first detectable in the 13th week of gestation (Fig. 3a). Some noticeable differences existed between fetal and adult esophageal Merkel cells: subepithelial Merkel cells, albeit very rare, could be found only in fetal specimens (Fig. 3b). They first occurred in the 20th

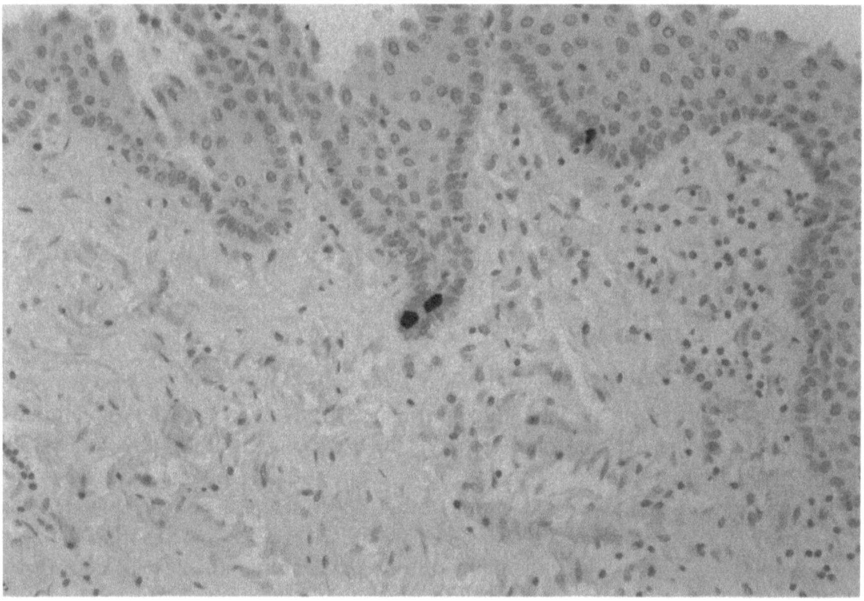

Fig. 1. Adult esophagus. Scattered roundish Merkel cells at the tips of the epithelial basal cell layer (CK 20 immunohistochemistry)

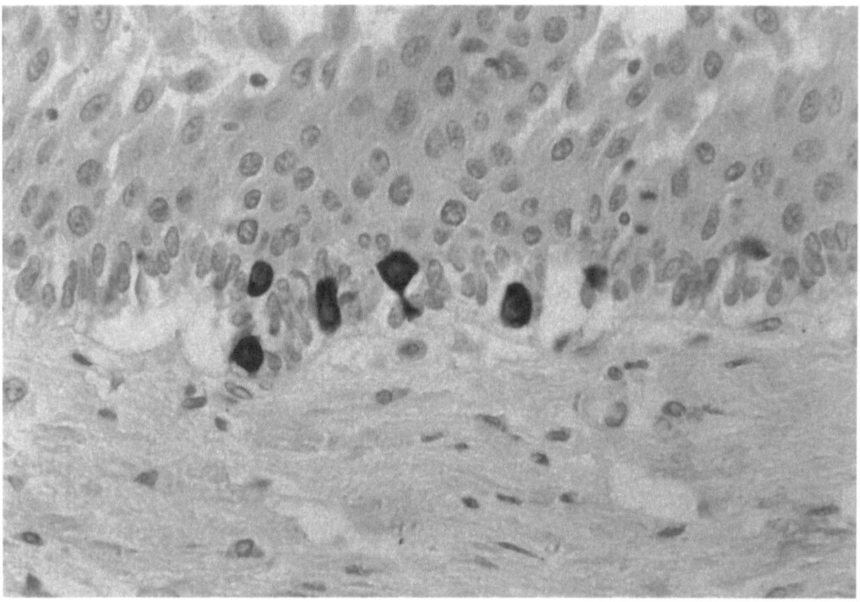

Fig. 2. Adult esophagus. Some grouped, mostly roundish Merkel cells. One Merkel cell has a clearly discernible dendrite (CK 20 immunohistochemistry)

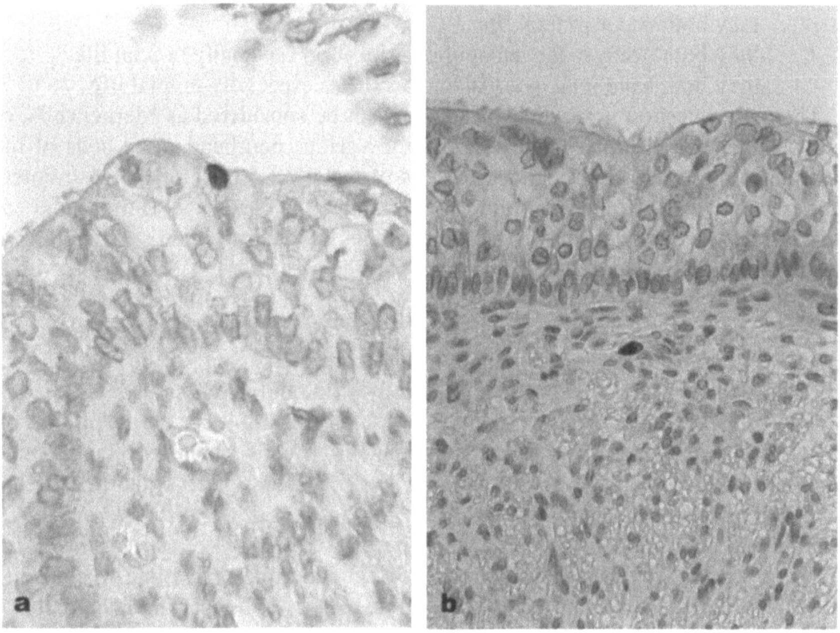

Fig. 3. a Fetal esophagus, 13th week. A single roundish Merkel cell at the surface of the still partially ciliated esophageal epithelium (CK 20 immunohistochemistry).**b** Fetal esophagus, 20th week. A single oval Merkel cell without dendrites situated in the subepithelial stroma (CK 20 immunohistochemistry)

week of gestation. In the fetal esophagi the epithelial Merkel cells were often arranged in small clusters (Fig. 4). In addition, the epithelial Merkel cells often possessed long dendrites, also well known in epidermal fetal Merkel cells (Fig. 5). Finally, the epithelial Merkel cells were often situated in the upper and even uppermost parts of the epithelium. This characteristic feature was especially striking in the esophagi of younger fetuses (see Fig. 3a). With advancing gestational age, the Merkel cells were located more and more in the lower parts of the epithelium (see Figs. 4, 5).

Discussion

The neuroendocrine cells in the esophagus demonstrated in our present study have many features in common with the neuroendocrine cells of the skin known as Merkel cells:

- they have a comparable immunoprofile
- they are in a comparable location as single cells lying in the basal layers of the epithelium
- they have a comparable density in the epithelium

- they both occur in fetal life
- they both occur in the subepithelial location especially in fetal life
- they both have long cytoplasmic dendrites especially in fetal life

The main question is whether these cells can be considered as Merkel cells, or if they must be subsumed simply under the various neuroendocrine cells of the gastrointestinal tract. Neuroendocrine cells of the several parts of the gastrointestinal tract are well known after all (Falkmer and Wilander 1995). Harmse and coworkers also investigated the esophageal probes in electron microscopy (Harmse et al. 1999). They found all the characteristic features of the Merkel cells. Moreover, the neuroendocrine cells of several parts of the gastrointestinal tract have been characterized in detail in electron microscopy. They have a different appearance and none of them is comparable in electron microscopy to the Merkel cell. Above all, they have different granules and they do not have desmosomes (Kaduk and Barth 1978; Falkmer and Wilander 1995). Therefore, it can be stated that on the electron microscopic level, the neuroendocrine cells of the esophagus have more in common with the Merkel cells of the skin than with the neuroendocrine cells of the remaining gastrointestinal tract.

Another reservation in accepting the neuroedocrine cells of the esophagus as Merkel cells could be that the esophagus is of endodermal derivation, whereas the skin is ectoderm. However, Merkel cells in an organ of endodermal location have already been described, namely in the tongue (Lacour et al. 1991). Therefore, the endodermal derivation of the esophagus cannot now be passed as an argument against the acceptance of esophageal Merkel cells.

In an important contrast to the Merkel cells of the skin, we never detected innervated Merkel cells in the esophagus. Interestingly, we found a great interindividual variation in the amount of esophageal Merkel cells. Often we could not detect any Merkel cells in the whole esophagus, whereas other specimens had a considerable amount of them. This observation raises some questions concerning the function of the esophageal Merkel cell. It can be assumed that the esophageal Merkel cells, like those in the skin, have neuroendocrine functions because they are likewise immunoreactive to neuroendocrine substances. However, these functions may not be essential, and Merkel cells in the esophagus possibly are a phylogenetic rudiment which in some humans does not exist any more. Therefore, it is tempting to speculate if the esophagus in lower mammals and other vertebrates has some special sensoric properties which are no longer expressed in men, and if in conjunction with these properties, the esophageal Merkel cells diminished, and therefore, do not show innervation any more. Consequently, future studies should look for Merkel cells and their possible innervation in various species.

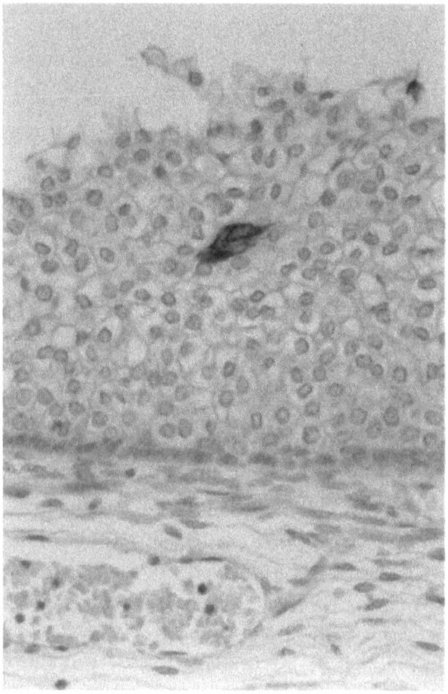

Fig. 4. Fetal esophagus, 26th week. A small cluster of dendritic Merkel cells in the mid-portion of the epithelium (CK 20 immunohistochemistry)

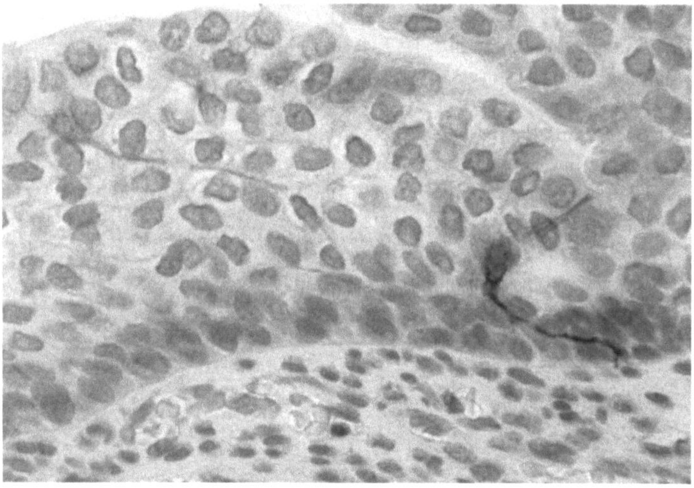

Fig. 5. Fetal esophagus, 28th week. An intensely stained Merkel cell with two very long dendrites on the right and a weaker stained Merkel cell with one long dendrite on the left in mid- and lower portions of the epithelium (CK 20 immunohistochemistry)

References

Botta MC, Ambu R, Liguori C, van Eyken P, Pisanu A, Cabras A, Hofler II, Werner M, Faa G (2001) CK 20 expression in the gastrointestinal tract of the embryo and fetus. Pathologica 93:640–644

Falkmer S, Wilander E (1995) The endocrine cell population. In: Whitehead R (ed) Gastrointestinal and oesophageal pathology, 2nd edn. Churchill Livingstone, New York, pp 63–71

Fetisoff F, Arbeille B, de Muret A, Heitzman A, Lansac J (1990) Cellules endocrines et épithéliums malpighiens. Cellules de type Merkel? Ann Pathol 10:323–330

Harmse JL, Carey FA, Baird AR, Craig SR, Christie KN, Hopwood D, Lucocq J (1999) Merkel cells in the human oesophagus. J Pathol 189:176–179

Hartschuh W, Weihe E, Reinecke M (1986) The Merkel cell. In: Bereiter-Hahn J, Matoltsy, AG, Richards KS (eds) Biology of the integument, vol 2 vertebrates. The Merkel cell. Springer, Berlin Heidelberg New York, pp 605–620

Hilliges M, Astback J, Wang L, Arvidson K, Johansson O (1996) Protein gene product 9.5 immunoreactive nerves and cells in human oral mucosa. Anat Rec 25:621–632

Hsu SM, Soban E (1982) Color modification of diaminobenzidine (DAB) precipitation by metallic ions and its application for double immunohistochemistry. J Histochem Cytochem 30:1079–1082

Hsu SM, Raine L, Fanger H (1981) Use of avidin–biotin–peroxidase complex (ABC) in immunoperoxidase techniques. J Histochem Cytochem 29:577–580

Kaduk B, Barth H (1978) Die Orthologie der endokrinen Zellen im distalen Oesophagus. Virchows Arch A Pathol Anat Histol 377:311–328

Lacour JP, Dubois D, Pisani A, Ortonne JP (1991) Anatomical mapping of Merkel cells in normal human adult epidermis. Br J Dermatol 125:535–542

Moll I, Bladt U, Jung EG (1990) Presence of Merkel cells in sun-exposed and not sun-exposed skin: a quantitative study. Arch Dermatol Res 282:213–216

Moll I, Kuhn C, Moll R (1995) Cytokeratin 20 is a general marker of cutaneous Merkel cells while certain neuronal proteins are absent. J Invest Dermatol 104:910–915

Tachibana T (1995) The Merkel cell: recent findings and unresolved problems. Arch Histol Cytol 58:379–396

Tateishi R, Taniguchi H, Wada A, Horai T, Taniguchi K (1974) Argyrophil cells and melanocytes in esophageal mucosa. Arch Pathol 98:87–89

Cellular Localization of Prepro-Orexin and Orexin Receptors (Ox1R and Ox2R) in Merkel Cells

A. Beiras-Fernández [1], M. Blanco[1], T. García-Caballero[1], R. Gallego[1], C. Diéguez[2] and A. Beiras[1]

[1]Department of Morphological Sciences, Faculty of Medicine, University of Santiago de Compostela, Spain
[2]Department of Physiology, Faculty of Medicine, University of Santiago de Compostela, Spain

Summary

Orexins (A and B) were first demonstrated in neurons of the lateral hypothalamus, both peptides derived from a common precursor called prepro-orexin. Orexins act through previously orphan G-protein receptors named orexin receptor 1 (Ox1R) and orexin receptor 2 (Ox2R). Up to the present, various neuropeptides have been detected in neuroendocrine cells of the skin (Merkel cells). The aim of our study was to investigate the presence of orexin receptors and prepro-orexins in Merkel cells. Immunohistochemical techniques were performed to detect prepro-orexin and orexin receptors (Ox1R and Ox2R) in human and porcine Merkel cells. Prepro-orexin was expressed in the cytoplasm of Merkel cells of pig snout skin and human fingertip. Immunoreactivity for prepro-orexin was more intense in the mature side of the Merkel cell. Epidermal nerve terminals associated with Merkel cells and dermal nerve fibers showed no immunostaining. Orexin receptors were also demonstrated in the Merkel cells of pig snout skin. Further studies must be followed to ascertain the role that orexins play in cutaneous neuroendocrine cells.

Introduction

Orexins (A and B) are peptides of different molecular weight discovered in the hypothalamus that arise by proteolysis from a 130-amino-acid common precursor called prepro-orexin (De Lecea et al. 1998). Orexins have been fundamentally described in cells from the lateral hypothalamus, also in dorsal and perifornical areas (Elias et al. 1998; Peyron et al. 1998) and both central and lateral medullar areas (Van den Pol 1999). Orexins have until now been involved in modulation of food intake (Sakurai 2002), macrophage function (Ichinose et al. 1998), stimulation of gastric acid secretion (Takahashi et al. 1999) and regulation of blood pressure, body temperature, neuroendocrine system and the sleep-wake cycle (Van den Pol et al. 1998; Date et al. 1999; Taheri et al. 2002). Merkel cells are neuroendocrine cells located in the basal layer of the epidermis and epithelium of oral mucosa as well as in the external root sheath of hair follicles (Beiras et al. 1987). Merkel cells express different neuropeptides, as shown by immunohistochemical studies as revised by Tachibana (Tachibana 1995). Other immunohistochemical studies have shown that Merkel cells can express polypeptides (Moll et al. 1984; Saurat et al. 1984); desmosomal proteins (Ortonne and Darmon 1985) and neuroendocrine markers such as neuron-specific enolase (Gu et al. 1981; Zaccone 1986); chromogranin A (Ness et al. 1987) or synaptophysin (Ortonne et al. 1988).

The aim of the present study was to investigate the presence of Ox1r and Ox2r in human and porcine neuroendocrine (Merkel) cells as well as to investigate the occurrence of its common precursor prepro-orexin.

Materials and Methods

Tissue specimens from pigs (snout) and humans (fingertips, lip) were immersion-fixed in Bouin's fluid for 2 h, dehydrated with a graded ethanol series and embedded in paraffin. Sections 5 μm thick were cut perpendicularly to the cutaneous surface, mounted on 3-aminopropyl-triethoxilane-coated slides and dried overnight at 37 °C. Immunohistochemical studies were performed after deparaffination and surrounding of the sections by Sigmacote (Sigma, St Louis, USA) with the help of rabbit polyclonal antisera, anti-prepro-orexin, anti-Ox1R and anti-Ox2R (Alpha Diagnostic, San Antonio, USA).

Antigen retrieval was carried out by means of pressure cooker unmasking for 2 min in 0.01 M sodium citrate buffer, pH 6.0. The streptavidin-biotin complex (SABC) was employed as immunohistochemical staining procedure with diaminobenzidine (DAB, Merck) being used as chromogen. Controls for specificity of the immunohistochemistry were performed including: (1) incubation with the primary antibodies previously preadsorbed with immunogen peptides, i.e., Prepro-orexin, Ox1R and Ox2R (Alpha Diagnostics) 10 nmol/ml, overnight at 4 °C; (2) replacement of one of the solutions of the previous steps with PBS.

Results

Merkel cells were located in pig snout skin and human fingertip epidermis with the help of SNE and semithin sections (Fig. 1). Pig snout skin and human epidermis were employed as controls of the immunohistochemical techniques. No immunostaining was found for prepro-orexin, Ox1R and Ox2R when any of the steps were omitted or when PBS was substituted for the antisera.

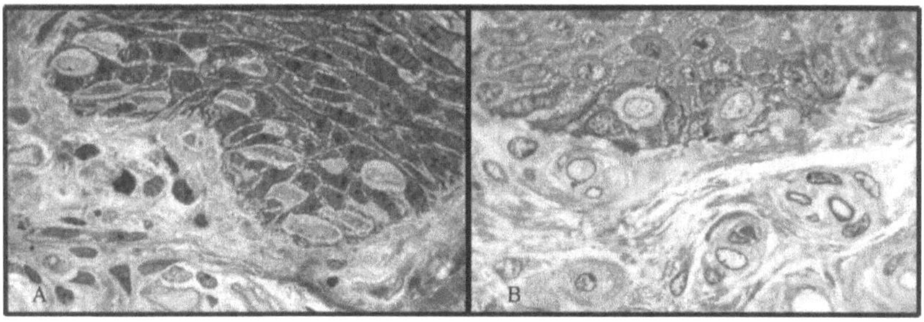

Fig. 1. Distribution of Merkel cells in pig snout skin (**A**) and in human epidermis (**B**). Semithin sections ×100

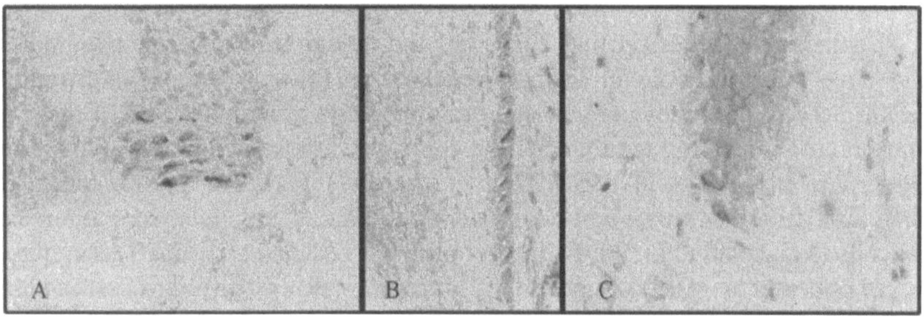

Fig. 2. Positive reaction for prepro-orexin in pig and human epidermis: **A** pig snout skin, **B** vibrissae, **C** human epidermis. Immunoreactivity is primarily localized in the mature side of the Merkel cells (×60)

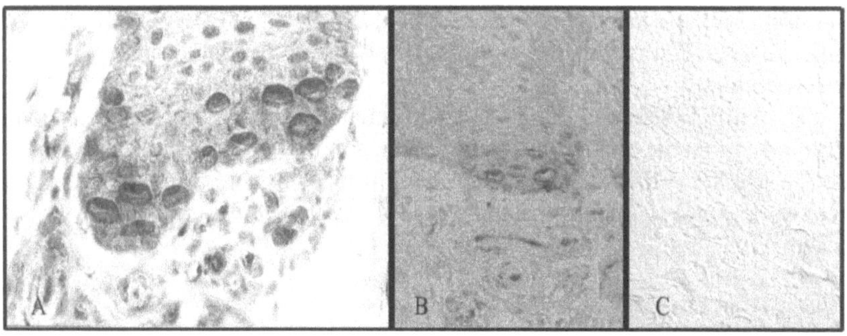

Fig. 3. Ox1R (**A**) and Ox2R (**B**) immunoreactivity is primarily observed in the superficial side of Merkel cells (×60). **C** Negative control (×20 Dic-Nomarsky)

Similar positive immunoreactivity was found for prepro-orexin both in human and porcine tissues (Fig. 2). Virtually all Merkel cells expressed primarily prepro-orexin, other epidermal cells being negative. Expression of prepro-orexin was more intense in the dermal side of the epidermal cells. In the vibrissae, Merkel cells positive for prepro-orexin were located in the outer layer of the external root sheath. Nuclei and epidermal nerve terminals associated with Merkel cells showed no positive reaction.

Positive immunoreactivity was found for OxR in porcine tissues. Immunostaining for both receptors was more intense in the superficial side of the cytoplasm of the cells (Fig. 3). Basal and lateral cytoplasmatic areas were also positive with weaker immunoreactivity. A positive reaction was found in the supranuclear area, where the synthesis and coating of neuroendocrine granules takes place. Perpendicular sections of superficial dermis that present this group of nerves showed a positive reaction to both receptors, being more intense in the case of Ox2R.

Discussion

Orexins are synthesized in neurons of the lateral and posterior hypothalamus and have a central role in feeding regulation. However, the demonstration of orexin nerve fibers projecting to the brain suggests a complex role of these peptides in autonomic and neuroendocrine control (Ichinose et al. 1998; Peyron et al. 1998; Van den Pol et al. 1998; Date et al. 1999). OxR and prepro-orexin have been also described in several neuroendocrine tissues and cells, e.g., adrenocortical cells (Mazzochi et al. 2001), human pituitary (Blanco et al. 2001), adrenal medulla (Lopez et al. 1999) or generally for prepro-orexin in different tissues (Jöhren et al. 2001).

This study describes for the first time the presence of orexin receptors and prepro-orexin, their common precursor in neuroendocrine Merkel cells. As reported in the results, immunoreactivity for prepro-orexin was shown in the dermal side of the epidermal cells, forming clusters at the base of the epidermis rete ridges while appearing on the outer layer of the external root sheath in the vibrissae. A positive response to OxR was found in the superficial side of the cytoplasm of the cells for both receptors.

Van den Pol (Van den Pol et al. 1998) related orexins to sensitivity regulation, especially in tactile and pain signaling, which is one of the main functions attributed to Merkel cells in the literature. Orexin receptors are also expressed in hair follicles and sweat glands and may have a relation to the autonomic regulative system.

Acknowledgements

We thank the expert technical assistance of D. Fernández-Roel. This work was supported by the Xunta de Galicia (grant no. PGIDT99PX120806B).

References

Beiras A, García-Caballero T, Espinosa J, Gallego R (1986) Staining of Merkel cells of pig snout epidermis using the uranaffin reaction. Morphometric analysis of neuroendocrine granules. Differentiation 32:89–92

Beiras A, García-Caballero T, Fernandez Redondo V, Gallego R (1987) Morphometric characterization of the human neuroendocrine Merkel cells. J Invest Dermatol 88:766–768

Blanco M, Lopez M, García-Caballero T, Gallego R, Vazquez-Boquete A, Morel G, Señaris R, Casanueva F, Dieguez C, Beiras A (2001) Cellular localization of orexin receptors in human pituitary. J Clin Endocrin Metabol 86:1616–1619

Date Y, Ueta Y, Yamashita H, Yamaguchi H, Matsukura S, Kangawa K, Sakurai T, Yanagisawa m, Nakazato M (1999) Orexins, orexigenic hypothalamic peptides, interact with autonomic, neuroendocrine and neuroregulatory systems. Proc Natl Acad Sci USA 96:748–753

De Lecea L, Kilduff TS, Peyron C, Gao XB, Foye P, Danielson PE, Fukuhara C, Battemberg E, Gautvik VT, Bartlett II FS, Frankel VN, Van den Pol AN, Bloom FE, Gautvik KM, Sutcliffe JG (1998) The hypocretins: hypothalamus-specific peptides with neuroexcitatory activity. Proc Natl Acad Sci USA 95:322–327

Dube MG, Kalra SP, Kalra P (1999) Food intake elicited by central administration of orexins/hypocretins: identification of hypothalamic sites of action. Brain Res 842:473–477

Gu J, Polak JM, Noorden SW, Pearse AGE, Marangos PZ, Azzopardi JG (1983) Immunostaining of neuron-specific enolase as a diagnostic tool for Merkel cell tumours. Cancer 52:1039–1043

Ichinose M, Asai M, Sawada M, Sasaki K, Oomura Y (1998) Induction of outward current by Orexin-B in mouse peritoneal macrophages. FEBS Lett 440:51–54

Jöhren O, Neidert SJ, Kummer M, Dendorfer A, Dominiak P (2001) Preproorexin and orexin receptors mRNAs are differentially expressed in peripheral tissues of male and female rats. Endocrinology 142:3324–3331

Lopez M, Señaris R, Gallego R, García-Caballero T, Lago F, Seoane L, Casanueva F, Dieguez C (1999) Orexin receptors are expressed in the adrenal medulla of the rat. Endocrinology 140:5991–5994

Mazzochi G, Malendowiez LK, Gottardo L, Aragona F, Nussdorfer GG (2001) Orexin-A stimulates cortisol secretion from human adrenocortical cells through activation of the adenylate cyclase-dependent signalling cascade. J Clin Endocrinol Metab 86:778–782

Moll R, Moll I, Franke WW (1984) Identification of Merkel cells in human skin by specific cytokeratin antibodies: changes of cell density and distribution in fetal and adult plantar epidermis. Differentiation 28:136–154

Ness KH, Morton TH, Dale BA (1987) Identification of Merkel cells in oral epithelium using antikeratin and antineuroendocrime monoclonal antibodies. J Dent Res 88:1154–1158

Ortonne JP, Darmon M (1985) Merkel cells express desmosomal proteins and cytokeratines. Acta Derm Venereol 65:161–164

Ortonne JP, Petchot-Bacque JP, Verrando P, Pisani A, Pautrat G, Bernerd F (1988) Normal Merkel cells express a synaptophysin-like immunoreactivity. Dermatologica 177:1–10

Peyron C, Tighe DK, Van den Pol AN, de Lecea L, Heller HC, Sutcliffe JC, Kilduff TS (1998) Neurons containing hypocretin (orexin) project to multiple neuronal system. J Neurosci 18:9996–10015

Piper DC, Upton N, Smith MI, Hunter AJ (2000) The novel brain neuropeptides, orexin-A, modulates the sleep-wake cycle in rats. Eur J Neurosci 12:726–730

Sakurai T (2002) Roles of orexins in regulation of feeding and wakefulness. Neuroreport 13:987–995

Sakurai T, Amemiya A, Ishii M, Matsuzaki I, Chemelli RM, Tanaka H, Williams SC, Richardson JA, Kozlowski GP, Wilson S, Arch JRS, Buckingham RE, Haynes AC, Carr SA, Annan RS, McNulty DE, Liu WS, Terret JA, Elshourbagy NA, Bergsma DJ, Yanasinawa M (1998) Orexins and orexins receptors: a family of hypothalamic neuropeptides and G-coupled protein receptors that regulate feeding behaviour. Cell 92:573–585

Saurat J, Merot Y, Didierjean L, Dahl D (1984) Normal rabbit Merkel cells do not express neurofilament proteins. J Invest Dermatol 82:641–642

Tachibana T (1995) The Merkel cell: recent findings and unresolved problems. Arch Histol Cytol 58:379–396

Taheri S, Zeitzer JM, Mignot E. (2002) The role of hypocretins (orexins) in sleep regulation and narcolepsy. Annu Rev Neurosci 25:283–313

Takahashi N, Okumura T, Yamada H, Kohgo Y (1999) Stimulation of gastric acid secretion by centrally administered Orexin-A in conscious rats. Biochem Biophys Res Commun 254:653–657

Van den Pol AN (1999) Hypothalamic hypocretin (orexin): robust inervation of the spinal cord. J Neurosci 19:3171–3182

Van den Pol AN, Gao XB, Obrietan K, Kilduff TS, Belousov AB (1998) Presynaptic and postsynaptic actions and modulation of neuroendocrine neurons by a new hypothalamic peptid: hypocretin/orexin. J Neurosci 18:7962–7971

Zaccone G (1986) Neuron-specific enolase and serotonin in the Merkel cells of conger-eel (Conger conger) epidermis. An immunohistochemical study. Histochemistry 85:29–34

Possible Functional Significance of Spatial Relationship Between Merkel Cells and Langerhans Cells in Human Hair Follicles

Yutaka Narisawa

Division of Dermatology, Department of Internal Medicine, Saga Medical School, Nabeshima 5-1-1, Saga City, 849-8501, Japan

Summary

Merkel cells and Langerhans cells share some similar characteristics such as the localization in hair follicles, close association with peripheral nerves and the localization of several neuropeptides. We found that Merkel cells and Langerhans cells were in contact with each other in the bulge area of human hair follicles under normal physiological conditions. This spatial relationship could imply that Merkel cells and Langerhans cells have a functional linkage within the specific site when stem cells are present or immature Langerhans cells are harbored. Merkel cells communicating with Langerhans cells in human hair follicles and Langerhans cells residing within the bulge should be considered. One explanation is that Merkel cells may support Langerhans cells that migrate into the basal layer of the bulge and then settle into this microenvironment. These mature or immature Langerhans cells, which are inhibited from circulating from the bulge to other regions, including lymph nodes and epidermis, may play some role in cross talk among stem cells, Merkel cells and/or peripheral nerves.

Introduction

Langerhans cells were first described in 1868 by Paul Langerhans, a German pathologist, and are dendritic cells principally situated in the epidermis. It was established by means of in vitro and in vivo experiments that these cells are bone marrow-derived leukocytes (Hosoi et al. 1993; Moresi and Horn 1997), which act as antigen-presenting cells, playing a key role in immune function (Merkel 1875; Narisawa et al. 1993, 1994). In recent years, Langerhans cells have been shown to express many neuropeptide receptors, and are closely connected with nerve fibers (Garcia-Cabellero et al. 1989). Langerhans cells belong to the neuroimmunocutaneous system (NICS; Tachibana et al. 1997).

Cutaneous neuromediators are synthesized by nerve fibers and Merkel cells and also Langerhans cells, keratinocytes, melanocytes and all immune cells (Tachibana et al. 1997). Merkel cells and Langerhans cells are generally considered to be independent from one another. Merkel cells and Langerhans cells share some similar characteristics such as the localization in the hair follicle, close association with peripheral nerves and the localization of several neuropeptides. We found an anatomic relationship between Merkel cells and Langerhans cells in the bulge area of human hair follicles. We will attempt to interpret the new finding that Merkel cells and Langerhans cells contact each other in the bulge area of human hair follicles under normal physiological conditions and consider the nature of Merkel cells.

Materials and Methods

Tissue. Normal scalp skin was embedded in paraffin and 8-μm serial transverse sections (200 sections) were prepared.

Preparation of Vellus Hair Follicles. The epidermal sheets with attached vellus hair follicles were prepared with EDTA. Hair follicles were observed three-dimensionally under a light microscope in the wet whole mount preparation.

Immunohistochemistry of Whole Mount Vellus Hair. The epidermal sheets were immunostained with monoclonal murine antibody CD1a (Novo Castra) as a marker for Langerhans cell and CAM5.2 (Beckton Dickinson) or CK20 (Progen) for Merkel cells.

Double Immunoperoxidase Staining. Serial transverse sections were immunostained using avidin-biotin-peroxidase complex (Vector Kit, Vector Laboratories, Burlingame, CA). A double immunoenzyme staining method with peroxidase (for cow S-100 protein) and alkaline phosphatase (for keratin) as labels was performed to study the topographical relationship between Langerhans cells and Merkel cells within the hair follicles, as previously described by us (Gu et al. 1981). Transverse sections were also stained with CD1a.

Double Immunofluorescence Labeling. The epidermal sheets were immunostained with monoclonal murine antibody CD1a as a marker for Langerhans cells, followed by goat anti-mouse immunoglobulin conjugated to FITC and then incubated with monoclonal murine antibody CK20 as a marker for Merkel cells, the secondary fluorescent-conjugated antibody was rhodamine-conjugated TRITC for CK20 labeling.

Morphometric Analysis. For numerical analyses of the relationship between Merkel cells and Langerhans cells, we used 8-μm serial transverse paraffinized sections (200 sections) of scalp skin. The numbers of ir-Merkel cells and Langerhans cells were counted. Merkel cells were registered when the nucleus-containing cell body was visible on the micrographs, but not when only fine cell processes or partial cross sections were included in the sections. On the other hand, the variably shaped cell bodies and dendrites of Langerhans cells were counted.

Results

Localization of Merkel Cells and Langerhans Cells in Hair Follicles

Transverse Sections. Merkel cells were preferentially situated in the outermost cell layer of the outer root sheath at the sebaceous gland level. On the other hand, Langerhans cells showing long dendritic processes were situated not only in the suprabasal layer, but also within the outermost cell layer of the outer root sheath, which is apparently different from the distribution pattern of the epidermis (Fig. 1).

Three-Dimensions. In the facial vellus hair follicles, Merkel cells were preferentially located within the enlarged portion below the sebaceous gland, which indicated the bulge area.

Langerhans cells were distributed in the facial vellus hair follicles. They were localized not only within the infundibulum, but also in the widened portion, presumably indicating the bulge.

Spatial Relationship Between Merkel Cells and Langerhans Cells Within the Human Hair Follicles

Transverse Sections. The double immunoenzyme technique using S-100 protein and CD1a antibodies on the transverse sections confirmed that S-100 protein-reactive follicular epithelial cells are indeed Langerhans cells, but not melanocytes.

At the sebaceous gland level of terminal hair follicles, several Merkel cells were in close contact with Langerhans cells on the outermost cell layer of the outer root sheath. Their relations were no contact with each cell body (Fig 2a) or direct contact (Fig. 2b), Merkel cell engulfed by Langerhans cell (Fig. 2c) and touching with each dendrite (Fig. 2d). There were numerous Langerhans cells without any association to Merkel cells in the outermost cell layer of the outer root sheath.

Three-Dimensions, Langerhans cells were widely distributed in vellus hair follicles, while Merkel cells were localized within the mid-portion of vellus hair follicles, presumably indicating the bulge area. In the bulge areas of vellus hair follicles, two different rendering approaches were visualized. Merkel cells and Langerhans cells were occasionally connected by cell bodies and branches.

Morphometric Data

The spatial relationship between Merkel cells and Langerhans cells was quantitatively shown. On Merkel cells, there was direct contact in 11/145 cells (7.6%), and 29/145 cells (20.0%) within one cell interval per two cells.

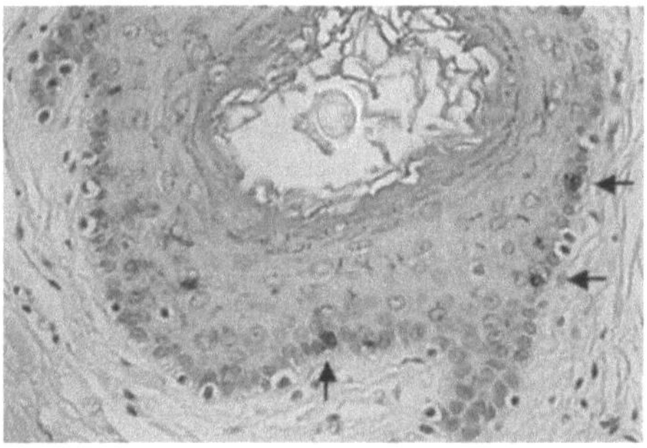

Fig. 1. Transverse section of human scalp skin. The localization of Langerhans cells in normal human skin. Langerhans cells showing variable dendritic processes are situated not only in the suprabasal layer, but also in the basal layer (*arrows*) of the outer root sheath. Paraffinized sections stained with CD1a (DAB); ×100

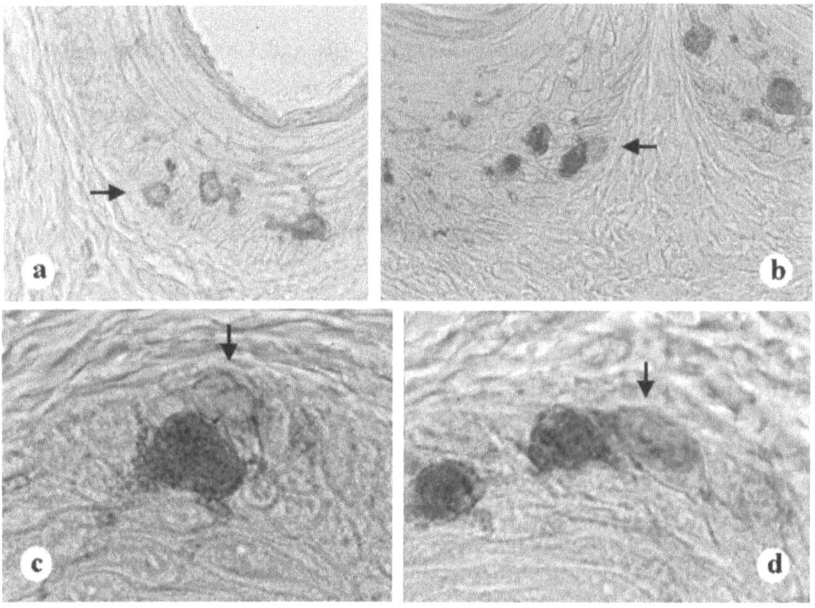

Fig. 2a–d. Transverse sections of normal human scalp skin. The special relationship between CAM5.2-reactive Merkel cells (*arrows*) and S-100 protein-reactive Langerhans cells is clearly shown in two dimensions. Note the variations such as a Merkel cell free from a Langerhans cell (**a**), direct contact (**b**), Merkel cell engulfed by Langerhans cell with long dendrites (**c**) and touching each dendrite (**d**). CAM5.2/alkaline phosphatase/Fast red for Merkel cell. S-100 protein/peroxidase/diaminobenzidine for Langerhans cell. **a**, **b** ×200 and **c**, **d** ×300

Discussion

Follicular stem cells reside within the bulge where they are interpreted to be an anatomical landmark of the attachment site of the arrector pili muscles (Hartschuh et al. 1989). Recently, the bulge area of the human hair follicle has attracted attention. The bulge area was composed of both resident and circulating cells. Since we were highly interested in the bulge, we investigated the bulge of human hair follicles with immunohistochemistry and the scanning electron microscope and then described several new findings.

The close relationship between nerve fibers and Merkel cells has been demonstrated with electron microscope and immunolabeling techniques. Merkel cells can be divided into two groups: innervated and noninnervated cells (Karanth et al. 1991). The functional significance of Merkel cells expressing several kinds of neuropeptides is still unknown. We consider that Merkel cells may be composed of functional heterogeneous subpopulations.

However, the localization of epidermal Langerhans cells principally in the suprabasal layer is anticipated to play a role in capturing antigens. Langerhans cells have been documented in the follicular epithelium by Breathnach (Thomas et al. 1984) and Jimbow et al. (Staniek et al. 1996). The immunology of the hair follicle remains biologically intriguing. The anagen proximal hair follicle constitutes an area of immune privilege within the hair follicle immune system (Torii et al. 1997). The mechanism by which contact hypersensitivity is induced most effectively in the skin with telogen hair follicles is poorly understood, but it is not simply due to failure of Langerhans cell function or sensitization of T cells (Christoph et al. 2000). This suggests that hair follicle cycling influenced skin immunity.

The most prominent cells located in or around the hair follicle are Langerhans cells, CD4+ or CD8+ T cells, macrophages and mast cells (Christoph et al. 2000). Gilliam et al. (1998) have recently reported the fascinating observation that the distal outer root sheath of human hair follicle harbors a special, rather immature Langerhans cell population. The cell has no desmosomes attaching it to adjacent cells.

In the present study, it was demonstrated that Merkel cells and Langerhans cells preferentially resided within the bulge areas of human hair follicles and directly connected or approached each dendrite. Whether follicular Merkel cells and/or Langerhans cells are associated with peripheral nerves will require further investigation. Nevertheless, this spatial relationship could imply that Merkel cells and Langerhans cells have a functional linkage within the specific site where stem cells are present or immature Langerhans cells are harbored.

The preferential localization of Langerhans cells within the basal layer of the bulge area is implicated to assume another function besides cutaneous immune response. Bone marrow-derived Langerhans cells migrate into skin appendages and subsequently participate in the cutaneous immune response circulating between skin and lymph nodes. The mechanism that circulating Langerhans cells arrive and collect in specific anatomical sites has not been discussed in detail. Since the bulge area is far from the overlying epidermis and the opening of the hair canal, the explanation that Langerhans cells resided within the bulge to trap, process and present antigens in the skin immune system is not easily accepted. Moreover, some Langerhans cells of the bulge area are situated in the basal layer different from the epidermis. One explanation is that Merkel cells may support Langerhans cells migrating into the basal layer and then settling into this microenvironment of the bulge area. These mature or immature Langerhans cells, which are inhibited from circulating from the bulge area to other regions including lymph nodes and epidermis, may play some role in cross talk among stem cells, Merkel cells and peripheral nerves. Although our explanation lacks evidence, similar functions have been suggested in the induction and rearrangement of peripheral nerves and arrector pili muscles in human fetal and postnatal skin. Further analysis of various surface molecule receptors for certain cytokines, various integrins and adhesion molecules is necessary to verify a functional heterogeneity of Langerhans cells locating within the bulge.

Our results demonstrated cell-to-cell contacts between Merkel cells and Langerhans cells in human hair follicles, although the functional importance remains unknown. In Merkel cell biology, Merkel cells communicating with Langerhans cells in human hair follicles should be considered. Moreover, the functional significance of Langerhans cells, especially in the basal layer of the bulge area, should be discussed in relation to the immunity and biology of hair follicles.

References

Christoph T, Muller-Rover S, Audring H, Tobin DJ, Hermes B, Cotsarelis G, Ruckert R, Paus R (2000) The human hair follicle immune system: Br J Dermatol 142:862–873

Garcia-Cabellero TR, Gallego E, Roson M, Beiras FA (1989) Calcitonin gene-related peptide (CGRP) immunoreactivity in the neuroendocrine Merkel cells and nerve fibers of pig and human skin. Histochemistry 92:127–132

Gilliam AC, Kremer IB, Yoshida Y, Stevens SR, Tootell E, Teunissen MB, Hammerberg C, Cooper KD (1998) The human hair follicle: a reservoir of CD40+ B7-deficient Langerhans cells that repopulate epidermis after UVB exposure. J Invest Dermatol 110:422–427

Gu J, Polak JM, Tapia FJ, Marangos PJ, Pearse AGE (1981) Neuron-specific enolase in the Merkel cells of mammalian skin. Am J Pathol 104:63–68

Hartschuh W, Weihe E, Yanaihara N (1989) Immunohistochemical analysis of chromogranin A and multiple peptides in the mammalian Merkel cell: further evidence for its paraneuronal function. Arch Histol Cytol 52:423–431

Hosoi J, Murphy GF, Egan CF, Lerner EA, Grabbe E, Asahina A, Granstein RD (1993) Regulation of Langerhans cell function by nerves containing calcitonin gene-related peptide. Nature 363:159–162

Karanth SS, Springall DR, Kuhn DM, Levene MM, Polak JM (1991) An immunohistochemical study of cutaneous innervation and the distribution of neuropeptides and protein gene product 9.5 in man and commonly employed laboratory animals. Am J Anal 191:369–383

Merkel F (1875) Tastzellen und Tastkörperchen bei den Hausthieren und beim Menschen. Arch Mikrosk Anat 11:636–652

Moresi JM, Horn TD (1997) Distribution of Langerhans cells in human hair follicle. J Cutan Pathol 24:636–640

Narisawa Y, Hashimoto K, Kohda H (1993) Epithelial skirt and bulge of human facial vellus hair follicles and associated Merkel cell-nerve complex. Arch Dermatol Res 285:269–277

Narisawa Y, Hashimoto K, Kohda H (1994) Merkel cells of the terminal hair follicle of the adult human scalp. J Invest Dermatol 102:506–510

Staniek V, Misery L, Peguet-Navarro J (1996) Expression of gastrin-releasing peptide receptor in human skin. Acta Dermatol Venereol 76:282–286

Tachibana T, Yamamoto H, Takahashi N, Kamegai T, Shibanai S, Iseki H, Nawa T (1997) Polymorphism of Merkel cells in the rodent palatine mucosa; immunohistochemical and ultrastructural studies. Arch Histol Cytol 60:379–389

Thomas JA, Biggerstaff M, Sloane JP, Easton DF (1984) Immunological analysis of regional variations of epidermal Langerhans cells in normal human skin. Histochem J 16:507–519

Torii H, Yan Z, Hosoi J, Granstein RD (1997) Expression of neurotrophic factors and neuropeptide receptors by Langerhans cells and the Langerhans cell-like cell line XS52: further support for a functional relationship between Langerhans cells and epidermal nerves. J Invest Dermatol 109:586–591

Neurobiology of the Murine Pelage Hair Follicle

E.M.J. Peters[1], V.A. Botchkarev[2], I. Moll[3], P. Arck[1] and R. Paus[3]

[1]Psychoneuroimmunology, Biomedical Research Center, Charité Campus Virchow Clinic, Berlin, Germany

[2]Department of Dermatology, Boston University School of Medicine, Boston, Massachusetts, USA

[3]Department of Dermatology, University Hospital Eppendorf, Hamburg, Germany

Summary

Exploitation of the murine pelage hair follicle as a model system that allows one to dissect and experimentally manipulate cutaneous neurobiology has generated a plethora of new insights into the roles of neuropeptides and neurotransmitters in the growth control of epithelial appendages in situ. Cross talk between epithelial-mesenchymal and neuronal cutaneous elements begins with the first development of skin and skin appendages. Development of cutaneous innervation of murine back skin hair follicles follows a strictly regulated spatio-temporal pattern. Cutaneous neural networks contain, e.g., neuropeptides such as calcitonin gene-related peptide (CGRP) and substance P (SP) in distinct distribution patterns. Compared to nontylotrich pelage hair follicles, tylotrich pelage hair follicles receive denser innervation and develop a lateral disc consisting of Merkel cells and SP-immunoreactive nerve fibers. Analysis of murine back skin innervation during the hair cycle has challenged the dogma that peripheral innervation only changes under pathological conditions such as injury. Most intriguingly, during anagen-development in murine back skin, SP-immunoreactive nerve fiber numbers fluctuate profoundly and transiently. Moreover, these fluctuations in nerve fiber numbers are accompanied by fluctuations in mast cell–nerve fiber contacts and in Merkel cell numbers. In organ-cultured early anagen back skin treated with SP, anagen progression can be promoted. These striking, hair cycle-dependent fluctuations in cutaneous innervation suggest trophic properties of piloneural interactions during hair cycle-associated skin remodeling and offer attractive, yet hitherto underexplored new targets for pharmacological intervention in the management of hair growth disorders.

Introduction

Are There "Trophic" Functions for Nerve Fibers in the Skin? The Neuro-"Trophic" Hypothesis

Almost 50 years have past since it first became evident that cutaneous innervation serves more than the classical effector and affector functions in skin (Kobayashi et al. 1958; Toksu 1970; Paus et al. 1997; Peters et al. 2001a). For example, we know now that peripheral innervation plays an important role in immunological processes in skin (Pincelli et al. 1993) and also covers a number of functions that can be summarized as "trophic". In the fetus or neonate, cutaneous innervation is essential in epithelial appendage development such as the hair follicle (Jones and Munger 1987; Maggi et al. 1987; Asada Kubota 1995), and in adult skin, wound healing is closely related to re-innervation of the insured skin and frequently associated with hair growth induction (Martin 1997; own, unpubl. obs.). However, cutaneous innervation has long been regarded as rather static and inflexible in regularly occurring tissue remodeling events as they take place during the hair cycle.

Many times throughout mammalian life, the hair follicle undergoes profound changes which have been described in detail in the C57BL/6 mouse (Moll et al. 1996; Botchkarev et al. 1997a; Mecklenburg et al. 2000; Müller-Röver et al. 2001). While growing (anagen), keratinocytes proliferate rapidly. Tylotrich and nontylotrich pelage hair follicles are densely innervated and accompanied by a rich vasculature, and Merkel discs, which are associated with the tylotrich pelage hair follicles only, contain many Merkel cells. This phase is followed by a regression phase (catagen) when the whole organ involutes. It challenged a dogma when it was shown for the first time that this process is associated with profound changes in cutaneous and hair follicle innervation (Moll et al. 1996; Botchkarev et al. 1997a,b; Peters et al. 2001a). The density of follicular and inter-follicular innervation decreases, as does the number of contacts between nerve fibers and mast cells and reduced numbers of Merkel cells are detectable in the Merkel discs. Ultimately, the hair follicle reaches a state of relative quiescence (telogen) and innervation density is low. Mast cell-nerve fiber contacts and Merkel cells are only detectable in very low numbers at this stage. These observations are suggestive for auto- and paracrine cascades of events involving neurogenic factors that exert trophic properties in the remodeling of the pilo-neural complex, which we want to look at in more detail throughout this article.

Morphogenesis of Hair Follicles and Cutaneous Innervation Are Closely Related

The "trophic" cross talk between epithelial-mesenchymal and neuronal cutaneous elements begins with the first development of skin and skin appendages (Jones and Munger 1987; Karanth et al. 1991; Asada Kubota 1995). As in other mammalian species (Mosconi and Rice 1993), the development and differentiation of cutaneous innervation and hair follicles progress in strictly correlated sequential waves in the back skin of C57BL/6 mice (Peters et al. 2002; Fig. 1). Tylotrich and later

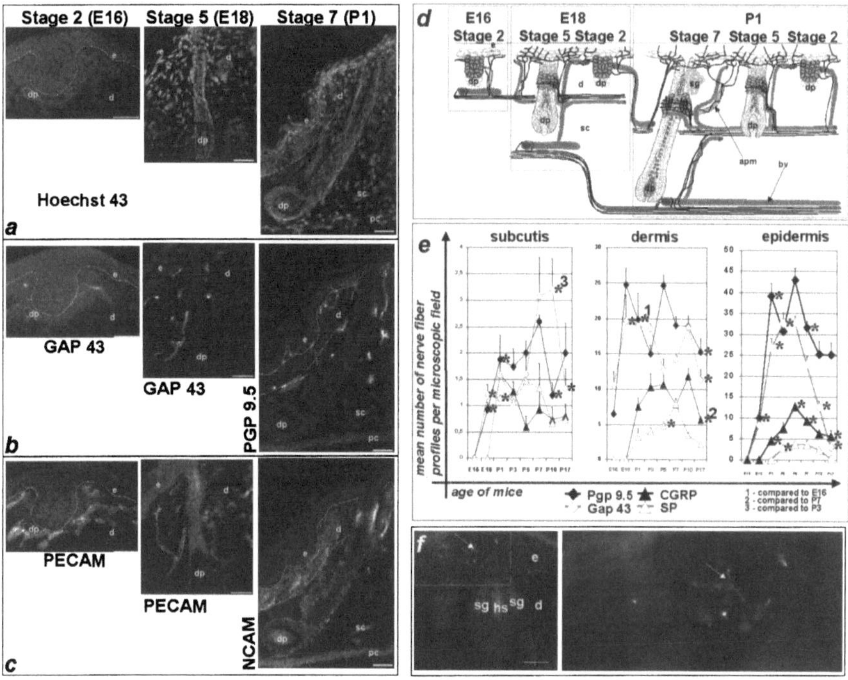

Fig. 1a–f. Development of nerve fibers in murine back skin during hair follicle morphogenesis. **a** Cell nuclei, Hoechst 43, **b** nerve fibers, **c** blood vessels – PECAM, neural adhesion – NCAM: representative images of hair follicles that develop into tylotrich pelage hair follicles. *Dotted lines* indicate basement membranes. **d** Schematic drawing summarizing cutaneous development: tylotrich pelage hair follicles (*E16*, stage 2, *E18*, stage 5, *P1*, stage 7) and nontylotrich pelage hair follicles (*E18*, stage 2, *P1*, stages 2 and 5). *Black lines* indicate developing innervation. *Dark gray profiles* indicate developing vasculature. *Gray circles* indicate NCAM-immunoreactive keratinocytes (hair follicle epithelium of the isthmus and bulge region), fibroblasts (dermal papilla) and smooth muscles (arrector pili muscles). **e** Quantification of nerve fiber profiles in skin samples of C57BL/6 mice taken on the prenatal days 16 and 18 (E16 and E18) and the postnatal days 1, 3, 5, 7, 10, 17 (P1 ...). Please note the transiently dense innervation of the epidermis with SP-immunoreactive nerve fibers in early postnatal life and the decline during the first hair follicle regression (P17). Significances have been tested by the Mann-Whitney-U test for unpaired samples and always refer to the previous day if not otherwise indicated (* $p<0.05$; ** $p<0.01$). Means are given ±SEM. **f** On day 3 of postnatal life, SP-immunoreactivity labels epidermal nerve fibers (*arrow* and *insert*) near a hair follicle with two sebaceous glands (=tylotrichen pelage hair follicle) in a cluster-like fashion, suggesting innervation of a Merkel disc as is commonly located in this position. *apm* Arrector pili muscle, *d* dermis, *dp* dermal papilla, *e* epidermis, *hs* hair shaft, *pc* panniculus carnosus, *sc* subcutis, *sg* sebaceous gland. *Scale bars* 50 μm

several waves of nontylotrich pelage hair follicles develop in the back skin together with their innervation (Mann 1962; Peters et al. 2002), so that every developing hair follicle of a given developmental stage is accompanied by nerve fibers

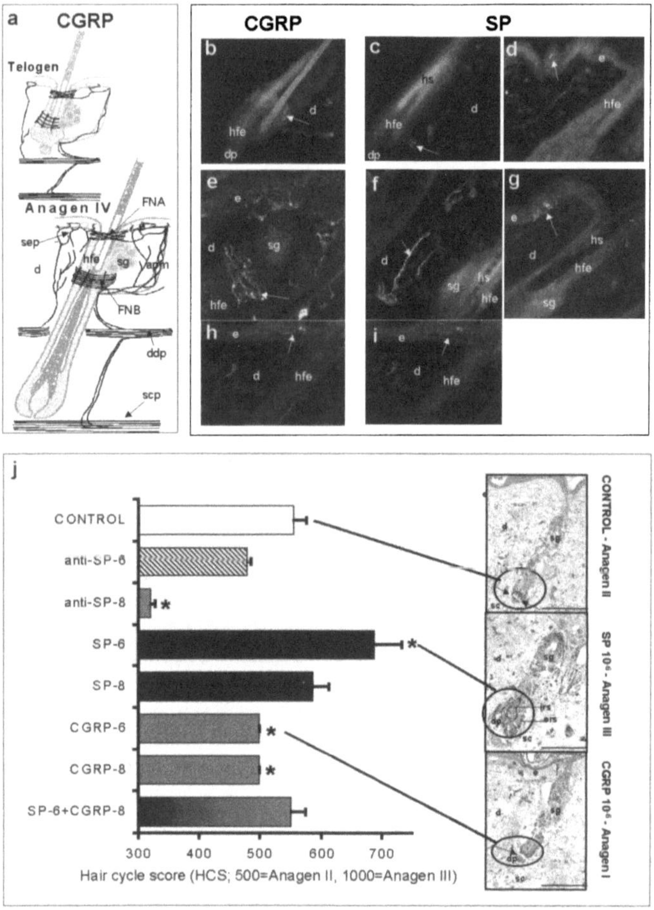

Fig. 2a–j. Sensory murine pelage hair follicle innervation and hair growth. **a** Schematic representations of a telogen and an early anagen hair follicle innervated by CGRP-immunoreactive nerve fibers (*black lines*). Single nerve fibers protrude from neural plexi (*sep*, *ddp*, *scp*) to innervate hair follicles, the arrector pili muscle, the epidermis, the dermis, and blood vessel (not shown). Hair follicle innervation is organized in two networks: follicular (neural) network A (*FNA*) and B (*FNB*). Nerve fiber density and complexity increase during the early anagen phase **b** (*CGRP*) and **c** (*SP*): telogen hair follicle with some immunoreactive nerve fibers in the FNB (*arrow* in **b**) and in a nerve fiber bundle in the deep dermal plexus passing by the bulge region (*arrow* in **c**). **d** Faint SP-immunoreactivity is located in the epidermis near a telogen tylotrich hair follicle most likely innervating a Merkel disc (*arrow*). **e** An anagen IV hair follicle with an FNB densely innervated by CGRP-immunoreactive nerve fibers (*arrow*). **f** Single SP-immunoreactive nerve fibers extend from the deep dermal plexus into the inter-follicular dermis (*arrow*) between anagen hair follicles. **g** A dense cluster of SP-immunoreactive nerve fibers innervates the epidermis near an anagen tylotrich hair follicle most likely innervating a Merkel disc (*arrow*). **h** and **i** CGRP-immunoreactive (**h**) and SP-immunoreactive (**i**) nerve fibers co-localize in nerve fibers innervating the epidermis close to a tylotrich hair follicle (*arrows*). **j** 4-mm punch →

matured in relation to that stage, not in relation to the maturity of the developing mammal.

Cutaneous innervation is densest during the first postnatal days (Fig. 1) and in the epidermis begins to decrease with the first penetration of the epidermis by developing hair shafts (Fig. 1), indicating the necessity for high epidermal sensitivity in the naked newborn skin. Neuropeptides become detectable in the developing mouse back skin nerve fibers around and after birth and again this follows a spatio-temporal pattern: first sensory neuropeptides such as calcitonine gene-related peptide (CGRP) and substance P (SP; Terenghi et al. 1993; Peters et al. 2002), then autonomic secretory neuropeptides such as vasoactive intestinal peptide (VIP) and peptide histidine methionin (PHM; Terenghi et al. 1993; Peters et al. 2001a) and last autonomic motor neuropeptides such as neuropeptide Y (NPY; Schotzinger and Landis 1990; Terenghi et al. 1993; Peters et al. 2001a). Interestingly, SP-immunoreactivity in the epidermis is concentrated to nerve fibers in small and defined fields close to tylotrich hair follicles as early as day 3 of postnatal life (Fig. 1).

Sensory Pelage Hair Follicle Innervation Fluctuates Hair Cycle Dependently

Cutaneous SP-content fluctuates hair cycle dependently with maximal SP-levels in early anagen, i.e., during the phase of rapid proliferation (Paus et al. 1994, 1995; Peters et al. 2001a). At the same time, during anagen-development, a significant increase in CGRP- and SP-immunoreactive nerve fiber numbers in dermis and subcutis and CGRP-immunoreactive nerve fibers in the FNB, around the hair follicle isthmus and bulge region between sebaceous gland and arrector pili muscle, can be observed (Peters et al. 2001a; Fig. 2). These observations suggest a role for SP and CGRP in anagen development, which prompted us to employ early anagen skin organ culture to investigate the effects of SP and CGRP on the growing C57BL/6 mouse back skin hair follicle. Interestingly, we found that CGRP had no effect despite its close relation to the hair follicle bulge region neighboring the

biopsies of C57BL/6 mouse skin, taken 3 days after anagen-induction by depilation, so that they contained only hair follicles in early anagen stages I–III, were cultured for 72 h in the presence of either SP (10^{-6} or 10^{-8} M), SP-antagonist (10^{-6} or 10^{-8} M), CGRP (10^{-6} or 10^{-8} M) or a combination of SP^{-6} and $CGRP^{-8}$ M). Quantitative histomorphometric analysis was performed and the number of follicles in telogen, anagen I, anagen II, and so forth documented. The hair cycle score (HCS) was calculated assigning a score of 0 to each anagen I hair follicle, 500 to anagen II, or 1000 to anagen III. Hair-cycle scores were added per sample and divided by the number of evaluated hair follicles. Significances have been tested by the Mann-Whitney-U test for unpaired samples and always refer to the previous day if not otherwise indicated (* $p<0.05$; ** $p<0.01$). Means are given ±SEM. *apm* Arrector pili muscle, *d* dermis, *ddp* deep dermal plexus, *dp* dermal papilla, *e* epidermis, *FNA* follicular (neural) network A, *FNB* follicular (neural) network B, *hfe* hair follicle epithelium, *hs* hair shaft, *scp* subcutaneous plexus, *sep* subepidermal plexus, *sg* sebaceous gland

FNB, containing one putative hair follicle stem cell population (Cotsarelis et al. 1999; Peters et al. 2001a). However, SP significantly promoted hair cycle progression in this model, and this effect could even be antagonized by CGRP (Peters et al. 2001a).

Conclusion

Given the close correlation of hair follicle and cutaneous innervation development, the number and complexity of perifollicular nerve fibers, the multitude of neuronal signaling molecules contained, and the hair cycle-associated fluctuations in perifollicular innervation, speculations on direct growth modulatory effects, e.g., on neighboring stem cells in the hair follicle bulge region are quickly made (Paus et al. 1997; Peters et al. 2002). Far from being able to give a complete review of all published data on this topic, this article has hopefully provided the reader with some evidence for a role for neuropeptides and neurotransmitters in hair growth control. In this context, it is interesting to note that SP-immunoreactive nerve fibers can only be detected in the epidermis close to tylotrich hair follicles, and these fibers increase in number during the hair growth phase. Hair cycle-dependent fluctuations in cutaneous innervation may thus reflect increased gene expression for neurotransmitters and neuropeptides, altered synthetic activity in the dorsal root ganglia or changes in peripheral digestion of released transmitters. In any case, they suggest trophic properties in hair cycle-associated skin remodeling.

Analysis of second messenger systems downstream of neurotransmitter and neuropeptide signalling will add to the understanding of piloneural hair growth control mechanisms (Peters et al. 2001b). Additional understanding can be derived from the analysis of related and interdependent systems such as the neurotrophin/neurotrophin-receptor system (Botchkareva et al. 2000). Also important will be the further analysis of interactions between the nervous system and cutaneous cells in close contact with neuronal elements such as Merkel cells or mast cells to increase our understanding of their interdependence. Future hair research in this field will focus on functionally relevant hair growth modulators such as SP and others in the human system, e.g., in the human adult anagen hair follicle culture system developed by Philpott (Peters et al. 2001b).

References

Asada Kubota M (1995) Inhibition of hair growth by subcutaneous injection of a sympathetic neurotoxin, 6-hydroxydopamine in neonatal mice. Anat Embryol Berl 191:407–414

Botchkarev VA, Eichmüller S, Johansson O, Paus R (1997a) Hair cycle-dependent plasticity of skin and hair follicle innervation in normal murine skin. J Comp Neurol 386:379–395

Botchkarev VA, Eichmüller S, Peters EM, Pietsch P, Johansson O, Maurer M, Paus R (1997b) A simple immunofluorescence technique for simultaneous visualization of mast cells and nerve fibers reveals selectivity and hair cycle-dependent changes in mast cell-nerve fiber contacts in murine skin. Arch Dermatol Res 289:292–302

Botchkareva NV, Botchkarev VA, Albers KM, Metz M, Paus R (2000) Distinct roles for nerve growth factor and brain-derived neurotrophic factor in controlling the rate of hair follicle morphogenesis. J Invest Dermatol 114:314–320

Cotsarelis G, Kaur P, Dhouailly D, Hengge U, Bickenbach J (1999) Epithelial stem cells in the skin: definition, markers, localization and functions. Exp Dermatol 8:80–88

Jones TE, Munger BL (1987) Neural modulation of cutaneous differentiation: epidermal hyperplasia and precocious hair development following partial neuralectomy in opossum pups. Neurosci Lett 79:6–10

Karanth SS, Springall DR, Kuhn DM, Levene MM, Polak JM (1991) An immunocytochemical study of cutaneous innervation and the distribution of neuropeptides and protein gene product 9.5 in man and commonly employed laboratory animals. Am J Anat 191:369–383

Kobayashi S, Okuyama F, Takagi K (1958) Experimental studies on the hemitrichosis and the nervous influences on the hair growth. Acta Neuroveg 18:169–190

Maggi CA, Borsini F, Santicioli P, Geppetti P, Abelli L, Evangelista S, Manzini S, Theodorsson-Norheim E, Somma V, Amenta F (1987) Cutaneous lesions in capsaicin-pretreated rats. A trophic role of capsaicin-sensitive afferents? Naunyn Schmiedebergs Arch Pharmacol 336:538–545

Mann SJ (1962) The prenatal formation of hair follicle types. Anat Rec 144:135–141

Martin P (1997) Wound healing – aiming for perfect skin regeneration. Science 276:75–81

Mecklenburg L, Tobin DJ, Müller-Röver S, Handjiski B, Wendt G, Peters EM, Pohl S, Moll I, Paus R (2000) Active hair growth (anagen) is associated with angiogenesis. J Invest Dermatol 114:909–916

Moll I, Paus R, Moll R (1996) Merkel cells in mouse skin: intermediate filament pattern, localization, and hair cycle-dependent density. J Invest Dermatol 106:281–286

Mosconi TM, Rice FL (1993) Sequential differentiation of sensory innervation in the mystacial pad of the ferret. J Comp Neurol 333:309–325

Müller-Röver S, Handjiski B, van der Veen C, Eichmüller S, Foitzik K, McKay IA, Stenn KS, Paus R (2001) A comprehensive guide for the accurate classification of murine hair follicles in distinct hair cycle stages. J Invest Dermatol 117:3–15

Parenti A, Amerini S, Ledda F, Maggi CA, Ziche M (1996) The tachykinin NK1 receptor mediates the migration-promoting effect of substance P on human skin fibroblasts in culture. Naunyn Schmiedebergs Arch Pharmacol 353:475–481

Paus R, Heinzelmann T, Schultz KD, Furkert J, Fechner K, Czarnetzki BM (1994) Hair growth induction by substance P. Lab Invest 71:134–140

Paus R, Heinzelmann T, Robicsek S, Czarnetzki BM, Maurer M (1995) Substance P stimulates murine epidermal keratinocyte proliferation and dermal mast cell degranulation in situ. Arch Dermatol Res 287:500–502

Paus R, Peters EM, Eichmüller S, Botchkarev VA (1997) Neural mechanisms of hair growth control. J Invest Dermatol Symp Proc 2:61–68

Peters EM, Botchkarev VA, Botchkareva NV, Tobin DJ, Paus R (2001a) Hair-cycle-associated remodeling of the peptidergic innervation of murine skin, and hair growth modulation by neuropeptides. J Invest Dermatol 116:236–245

Peters EM, Moll I, Alt V, Löffler D, Paus R (2001b) Choleratoxin but not isoproterenol inhibits human hair follicle growth in vitro. J Invest Dermatol 117:428 (abstract)

Peters EM, Botchkarev VA, Müller-Röver S, Moll I, Rice FL, Paus R (2002) Developmental timing of hair follicle and dorsal skin innervation in mice. J Comp Neurol 448:28–52

Pincelli C, Fantini F, Giannetti A (1993) Neuropeptides and skin inflammation. Dermatology 187:153–158

Schotzinger RJ, Landis SC (1990) Postnatal development of autonomic and sensory innervation of thoracic hairy skin in the rat. A histochemical, immunocytochemical, and radioenzymatic study. Cell Tissue Res 260:575–587

Terenghi G, Sundaresan M, Moscoso G, Polak JM (1993) Neuropeptides and a neuronal marker in cutaneous innervation during human foetal development. J Comp Neurol 328:595–603

Toksu EA (1970) Trophic influence of sensory nerves on hair follicles in mice. Plast Reconstr Surg 46:71–75

Development

Mammalian Merkel Cells Are Neural Crest Derivatives

M. Sieber-Blum[1], V. Szeder[1, 2], M. Grim[2], and Z. Halata[3]

[1]Department of Cell Biology, Neurobiology and Anatomy, Medical College of Wisconsin, Milwaukee, Wisconsin 53226, USA

[2]Institute of Anatomy, First Faculty of Medicine, Charles University Prague, 12800 Prague, Czech Republic

[3]Department of Functional Anatomy, University of Hamburg, Hamburg, 20246 Germany

Summary

We provide evidence for the neural crest origin of mammalian Merkel cells. Neural crest cells originate in the neural folds during early development of the vertebrate embryo. They delaminate from the dorsal aspect of the forming neural tube and emigrate to different locations, giving rise to a diverse array of structures in the adult organism. The dorsal neural tube, including neural crest cells, transiently express the protein, Wnt-1. This feature can be used to genetically mark neural crest cells and their derivatives. We have thus used the double transgenic Wnt1-cre/R26R mouse to determine the ontogenetic origin of mammalian Merkel cells. Merkel cells in the hair follicle epithelium of the whisker pad, interfollicular Merkel cells in touch domes and Merkel cells in the rete ridge express the neural crest-specific transgene, β-galactosidase. Our data thus indicate that mammalian Merkel cells are derived from the neural crest, resolving a long-standing controversy. Moreover, they suggest that Merkel cell carcinomas are not epithelial tumors, but belong to the neurocristopathies.

Introduction

The neural crest is a transient structure of the vertebrate embryo. Upon formation of the neural tube, cells of the neural folds delaminate and migrate via different pathways into the developing embryo. In the adult organism, neural crest cells give rise to a wide variety of cell types and tissue. They include all cells of the autonomic and enteric nervous systems, most primary sensory neurons and endocrine cells such as the adrenal medulla and the C-cells of the thyroid. In addition, neural crest cells form the cranial mesenchyme, which gives rise to facial skin, bone and connective tissue among other cranial structures (Le Douarin and Kalcheim 1999). As Merkel cells show many features of endocrine cells, it is conceivable that they also originate from the neural crest. Indeed, we have shown previously with the chick-quail transplantation method that avian Merkel cells are neural crest derivatives (Halata et al. 1990; Grim and Halata 2000a,b). However, the current prevalent opinion is that mammalian Merkel cells are derived from the epidermis. This notion is based on their epidermal location, their expression of cytokeratins, and skin transplantation experiments (Lyne and Hollis 1971; English et al. 1980; Moll et al. 1984, 1986). In contrast, our previous studies have shown that neural crest stem cells are among the migratory population of neural crest cells, and that they are also present in target locations, including the sympathetic ganglia, dorsal root ganglia and cardiac outflow tract (Duff et al. 1991; Ito and Sieber-Blum 1993; Richardson and Sieber-Blum 1993; Sieber-Blum et al. 1993). In vitro clonal analysis has shown further that the leading edge of the migrating neural crest cell population that migrates subectodermally within the dorsolateral pathway contains stem cells (Richardson and Sieber-Blum 1993; Sieber-Blum et al. 1993). The presence of stem cells in the migratory and post-migratory neural crest explains the extraordinary diverse progeny of this tissue. Moreover, it is thus conceivable that the neural crest also gives rise to Merkel cells.

During embryonic development, Wnt-1 is expressed only in the central nervous system and in the neural crest (for review, see Echelard et al. 1994). We used a two-component genetic system that takes advantage of the restricted Wnt-1 expression and indelibly marks neural crest cells and their derivatives. Wnt1-cre mice (Danielian et al. 1998) express the viral recombinase, cre, under the control of the Wnt-1 promoter. R26R mice are derivatives of the ROSA26 line (Friedrich and Soriano 1991). The latter is the result of a gene trap that expresses β-galactosidase ubiquitously under the control of an unidentified promoter (Soriano 1999). Thus, in ROSA26 mouse embryos, all cells express β-galactosidase. The R26R mouse has, in addition, a stop codon upstream of β-galactosidase that is flanked by loxP sites, recombination sites that are recognized by cre (Soriano 1999). Therefore, in Wnt1-cre/R26R double transgenic mice, neural crest cells and dorsal neural tube cells of brain and trunk axial levels and their derivatives express β-galactosidase indefinitely (Friedrich and Soriano 1991; Echelard et al. 1994; Danielian et al. 1998; Soriano 1999; Chai et al. 2000; Jiang et al. 2002).

Results

Mice were genotyped and stained for Xgal histochemical stain, TROMA-1 immunoreactivity and β-galactosidase immunoreactivity as described (Szeder et al. 2002). The Wnt1::lacZ mouse (Echelard et al. 1994; Jackson Laboratories) expresses β-galactosidase under the control of the Wnt1 promoter. In this mouse,

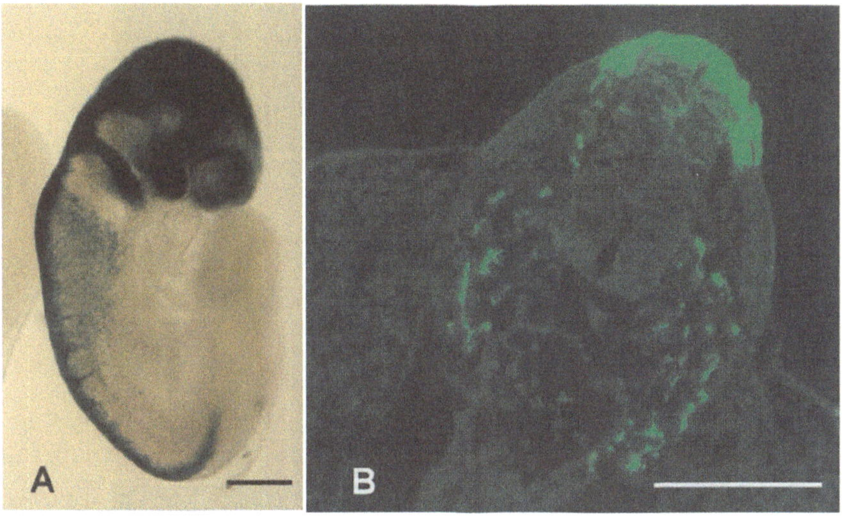

Fig. 1A, B. Wnt-1 expression in an E 9.5 Wnt-1::lacZ mouse embryo. In these transgenic mice, β-galactosidase is expressed upon activation of the Wnt-1 promoter. **A** Xgal histochemistry. Migrating neural crest cells in cranial and trunk axial levels are Xgal-positive. **B** Cross section through the trunk area of an E 9.5 Wnt-1::lacZ embryo, stained with anti-β-galactosidase antibodies (ALEXA488; *green* fluorescence). Wnt-1 expression is limited to the migratory and premigratory neural crest, and to the dorsal neural tube. *Calibration bars* **A** 500 μm, **B** 100 μm

neural crest cells transiently express β-galactosidase. In an embryonic day (E) 9.5 Wnt1::lacZ embryo, migrating Xgal-positive neural crest cells are evident in cranial and trunk areas (Fig. 1A), demonstrating that transgenic neural crest cells are formed and migrate normally. In cross sections, premigratory and migratory neural crest cells, as well as dorsal neural tube noncrest cells, strongly bind anti-β-galactosidase antibodies (Fig. 1B). In E 9.5 double transgenic Wnt1-cre/R26R mice, dorsal neural tube cells, premigratory neural crest cells and migratory neural crest cells express cre, as detected with an anti-cre antibody (Fig. 2A). At E 16.5, epidermal cells express the Wnt-3 and Wnt-5a genes (St.-Jacques et al. 1998; Millar et al. 1999; Fuchs et al. 2001; Reddy et al. 2001), but not Wnt-1. Wnt-1 expression during embryonic development is restricted to the central nervous system and the neural crest (reviewed by Echelard et al. 1994). As an important negative control, we have nevertheless confirmed the absence of Wnt-1 expression in E 16.5 epidermal cells. Neither the outer root sheet in whiskers (Fig. 2B, C), nor the surface epidermis (Fig. 2D, E) express cre at E 16.5. Due to protein turnover, the dermal cells, which are of neural crest origin, are also cre-negative, as the Wnt1 promoter is active transiently only (Fig. 2B–E), around E 9.5.

Merkel cells are visualized by TROMA-1 binding, a monoclonal antibody that recognizes cytokeratin 8 (Vielkind et al. 1995; Moll et al. 1996a; Fig. 3A, C). TROMA-1 positive Merkel cells (Fig. 3C) express β-galactosidase (Fig. 3B).

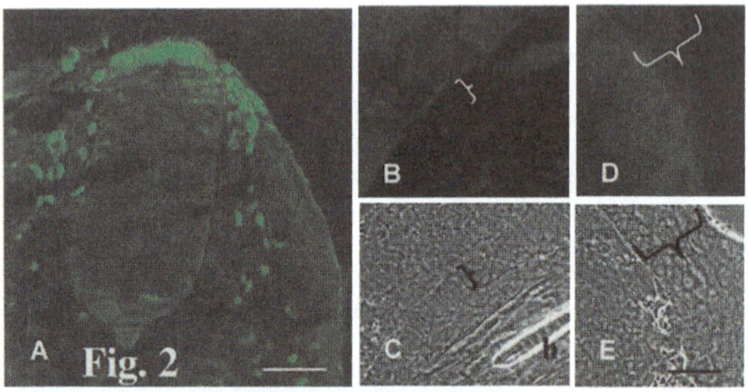

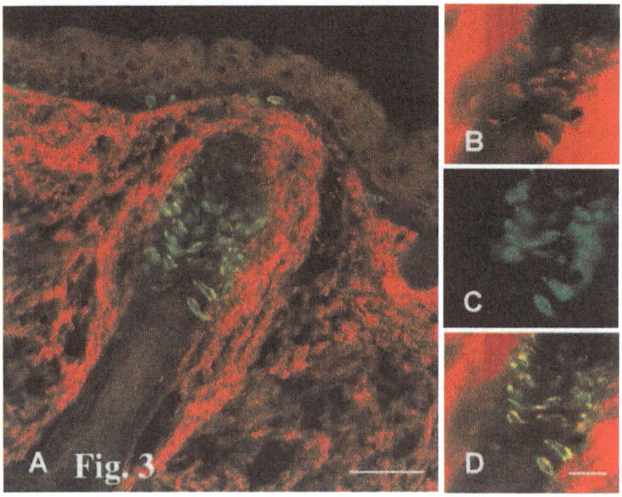

Fig. 2A–E. Cre expression in E 9.5 and E 6.5 Wnt1-cre/R26R embryos. **A** Cre immunoreactivity in a cross section of an E 9.5 Wnt1-cre/R26R embryo is limited to the neural crest and dorsal neural tube, as expected. **B** In E 16.5 whiskers, no cre-immunoreactivity is observed. *Bracket* denotes the outer root sheet, where Merkel cells are located. **C** Same area with phase contrast optics; *h* hair. **D** There is no cre-immunoreactivity in the surface epidermis (*bracket*). **E** Same area as **D** with phase contrast optics. The absence of cre-immunoreactivity in epidermal cells excludes the possibility of Wnt-1 expression in epidermal cells at that stage of development. The absence of cre-immunoreactivity in E 16.5 dermal cells and Merkel cells, both of which are of neural crest origin, is consistent with the transient expression of Wnt-1 around day E 9.5 and a subsequent loss of cre protein due to protein turnover. *Calibration bars* **A–E** 50 μm

Fig. 3A–D. β-galactosidase expression in the whisker pad of E 16.5 Wnt1-cre/R26R double transgenic mice. **A** Merged image of β-galactosidase (ALEXA 594-conjugated goat anti-rabbit IgG; *red* fluorescence) and TROMA-1 immunoreactivity (*fluorescein* fluorescence); overview of one whisker. **B** β-galactosidase immunoreactivity (ALEXA 594, *red* fluorescence); detail from **A**. **C** TROMA-1 immunoreactivity (*fluorescein* fluorescence) of same area as in **B**, same focal plane. **D** Merged images of **C** and **D**; β-galactosidase immunoreactivity co-localizes with TROMA-1 immunoreactivity (*yellow*). In control experiments where one of the primary antibodies or one of the secondary antibodies was deleted, there was no detectable staining (data not shown). *Calibration bars* **A** 50 μm, **B–D** 10 μm

Figure 3D shows a merged image with superimposed TROMA-1 (green) and β-galactosidase (red) stains. Areas of co-expression appear yellow. The co-localization of cytokeratin 8 and β-galactosidase provides conclusive evidence for the neural crest origin of Merkel cells (Szeder et al. 2002). β-galactosidase is expressed abundantly also in the surrounding dermal tissue of the developing whisker pad (Fig. 3A), as the cranial mesenchyme is of neural crest origin.

Discussion

We used a two-component genetic system to prove that Merkel cells are derived from the neural crest. The neural crest generates highly diverse progeny, including several types of endocrine cell that according to their morphology are closely related to Merkel cells (Le Douarin and Kalcheim 1999). Moreover, our previous studies have shown that avian Merkel cells are derived from the neural crest (Halata et al. 1990; Grim and Halata 2000a,b). In evolutionary terms, it is thus not surprising that mammalian and avian Merkel cells share their common origin. However, many studies have been geared towards proving that Merkel cells have an epidermal origin. First and foremost, mammalian Merkel cells are located in and integrated with the basal layer of the epidermis. In the whisker they are located in the outer root sheet, which is in direct continuum with the epidermis. Moreover, Merkel cells express several types of cytokeratin, including cytokeratin 8 (Moll et al. 1990; Vielkind et al. 1995). Thus, the potential epidermal nature of Merkel cells becomes plausible. However, in addition to the chick-quail transplantation experiments and our current report, there are other indications that contradict the notion of an epidermal origin. First, in mammalian fetal skin, Merkel cells are observed occasionally in the dermis (Hashimoto 1972; Halata 1981; Tachibana 1995). Second, Merkel cell carcinomas arise in the dermis, not the epidermis (Goessling et al. 2002). Interestingly, in rare cases Merkel cell carcinoma occurs in patients with neurofibromatosis, another neural crest disease (Antoniades et al. 1998). Merkel cell carcinomas also express Brn-3.0 (Leonard and Bell 1997), which is characteristic for neural crest-derived sensory neurons (Greenwood et al. 1999). Third, the low molecular weight cytokeratins that are expressed by Merkel cells are not characteristic for epidermis, but for simple epithelia (Kemler et al. 1981). Finally, Moll and collaborators have performed transplantation experiments by grafting human fetal skin from gestational weeks 8–11 to nude mice. In the graft, human Merkel cells developed. The authors have thus concluded that they are derived from the epidermis. The problem with this interpretation is that Merkel cell progenitors and neural crest-derived glia cells are already present in gestational weeks 6–11 human fetal skin (Moore and Munger 1989; Terenghi et al. 1993; Moll et al. 1996b). For this reason and because many target locations in the young embryo contain neural crest stem cells (Duff et al. 1991; Ito and Sieber-Blum 1993; Richardson and Sieber-Blum 1993; Sieber-Blum et al. 1993), it is likely that neural crest stem cells invade the skin early in embryonic development, and that they receive cues from the local microenvironment that direct them to differentiate along the Merkel cell lineage. In summary, we have provided conclusive evidence for the neural crest origin of mammalian Merkel cells, refuting the notion of their epidermal origin. Furthermore, our data suggest that Merkel cell carcinomas are not epithelial tumors, but that they belong to the neurocristopathies (Bolande 1974).

Acknowledgements

We thank A.P. McMahon, P. Soriano and H. Sucov for providing the Wnt1-cre and R26R mouse lines and J. Sanes for a generous gift of β-galactosidase antibodies. We thank Joan Ward, Eva Kluzáková and Brigitte Asmus for excellent technical assistance. The TROMA-1 antibody developed by P. Brulet and R. Kemler was purchased from the Developmental Studies Hybridoma Bank developed under the auspices of the NICHD and maintained by the University of Iowa, Department of Biological Sciences, Iowa City, IA 52242. This work was supported by USPHS grant NS38500 (MSB) and postdoctoral fellowship 1F05-11111-02 (VS) from the National Institute of Neurological Disorders and Stroke; by grants VZ 1111 00003-3G (MG) and LN 00A065 (MG) from the Ministry of Education of the Czech Republic and by DAAD (Deutscher Akademischer Austausch Dienst; ZH).

References

Antoniades K, Giannouli T, Kaisaridou D (1998) Merkel cell carcinoma in a patient with Recklinghausen neurofibromatosis. Int J Oral Maxillofac Surg 27:213–214

Bolande RP (1974) the Neurocristopathies: a unifying concept of disease arising in neural crest maldevelopment. Hum Pathol 5:409–429

Chai Y, Jiang X, Ito Y, Bringas P Jr, Han J, Rowitch DH, Soriano P, McMahon AP, Sucov HM (2000) Fate of the mammalian cranial neural crest during tooth and mandibular morphogenesis. Development 127:1671–1679

Danielian PS, Muccino D, Rowitch DH, Michael SK, McMahon AP (1998) Modification of gene activity in mouse embryos in utero by a tamoxifen-inducible form of Cre recombinase. Curr Biol 8:1323–1326

Duff RS, Langtimm CJ, Richardson MK, Sieber-Blum M (1991) In vitro clonal analysis of progenitor cell patterns in dorsal root and sympathetic ganglia of the quail embryo. Dev Biol 147:451–459

Echelard Y, Vassileva G, McMahon AP (1994) *Cis*-acting regulatory sequences governing Wnt-1 expression in the developing mouse CNS. Development 120:2213–2224

English KB, Burgess PR, Kavka-Van Norman D (1980) Development of rat Merkel cells. J Comp Neurol 194:475–496

Friedrich G, Soriano P (1991) Promoter traps in embryonic stem cells: a genetic screen to identify and mutate developmental genes in mice. Genes Dev 5:1513–1523

Fuchs E, Merrill BJ, Jamora C, DasGupta R (2001) At the roots of a never-ending cycle. Dev Cell 1:13–25

Goessling W, McKee PH, Mayer RJ (2002) Merkel cell carcinoma. J Clin Oncol 20:588–598

Greenwood AL, Turner EE, Anderson DJ (1999) Identification of dividing, determined sensory neuron precursors in the mammalian neural crest. Development 126:3545–3559

Grim M, Halata Z (2000a) Developmental origin of avian Merkel cells. Anat Embryol 202:401–410

Grim M, Halata Z (2000b) Developmental origin of Merkel cells in birds. In: Suzuki H, Ono T (eds) Merkel cells, Merkel cell carcinoma and neurobiology of the skin. Elsevier, Amsterdam, pp 23–32

Halata Z (1981) Postnatal development of sensory nerve endings in the hairless nose skin of the cat. Bibl Anat 19:210–235

Halata Z, Grim M, Christ B (1990) Origin of spinal cord meninges, sheaths of peripheral nerves, and cutaneous receptors including Merkel cells. An experimental and ultrastructural study with avian chimeras. Anat Embryol 182:529–537

Hashimoto K (1972) The ultrastructure of the skin of human embryos. X. Merkel tactile cells in the finger and nail. J Anat 111:99–120

Ito K, Sieber-Blum M (1993) Pluripotent and developmentally restricted neural-crest-derived cells in posterior visceral arches. Dev Biol 156:191–200

Jiang X, Iseki S, Maxson RE, Sucov HM, Morriss-Kay GM (2002) Tissue origins and interactions in the mammalian skull vault. Dev Biol 241:106–116

Kemler R, Brulet P, Schnebelen MT, Gaillard J, Jacob F (1981) Reactivity of monoclonal antibodies against intermediate filament proteins during embryonic development. J Embryol Exp Morphol 64:45–60

Le Douarin NM, Kalcheim C (eds) (1999) The neural crest. Cambridge University Press, Cambridge

Leonard JH, Bell JR (1997) Insights into the Merkel cell phenotype from Merkel cell carcinoma lines. Australas J Dermatol 38:S91–98

Lyne AG, Hollis DE (1971) Merkel cells in sheep epidermis during fetal development. J Ultrastruct Res 34:464–472

Millar SE, Willert K, Salinas PC, Roelink H, Nusse R, Sussman DJ, Barsh GS (1999) WNT signaling in the control of hair growth and structure. Dev Biol 207:133–149

Moll I, Moll R, Franke WW (1986) Formation of epidermal and dermal Merkel cells during human fetal skin development. J Invest Dermatol 87:779–787

Moll I, Lane AT, Franke WW, Moll R (1990) Intraepidermal formation of Merkel cells in xenografts of human fetal skin. J Invest Dermatol 94:359–364

Moll I, Paus R, Moll R (1996a) Merkel cells in mouse skin: intermediate filament pattern, localization, and hair cycle-dependent density. J Invest Dermatol 106:281–286

Moll I, Ziegler W, Schmelz M (1996b) Proliferative Merkel cells were not detected in human skin. Arch Dermatol Res 288:184–187

Moll R, Moll I, Franke WW (1984) Identification of Merkel cells in human skin by specific cytokeratin antibodies: changes of cell density and distribution in fetal and adult plantar epidermis. Differentiation 28:136–154

Moore SJ, Munger BL (1989) The early ontogeny of the afferent nerves and papillary ridges in human digital glabrous skin. Brain Res Dev Brain Res 48:119–141

Reddy S, Andl T, Bagasara A, Lu MM, Epstein DJ, Morrisey EE, Millar SE (2001) Characterization of Wnt gene expression in developing and postnatal hair follicles and identification of Wnt5a as a target of Sonic hedgehog in hair follicle morphogenesis. Mech Dev 107:69–82

Richardson MK, Sieber-Blum M (1993) Pluripotent neural crest cells in the developing skin of the quail embryo. Dev Biol 157:348–358

Sieber-Blum M, Ito K, Richardson MK, Langtimm CJ, Duff RS (1993) Distribution of pluripotent neural crest cells in the embryo and the role of brain-derived neurotrophic factor in the commitment to the primary sensory neuron lineage. J Neurobiol 24:173–184

Soriano P (1999) Generalized lacZ expression with the ROSA26 Cre reporter strain. Nat Genet 21:70–71

St-Jacques B, Dassule HR, Karavanova I, Botchkarev VA, Li J, Danielian PS, McMahon JA, Lewis PM, Paus R and McMahon AP (1998) Sonic hedgehog signaling is essential for hair development. Curr Biol 8:1058–1068

Szeder V, Grim M, Halata Z, Sieber-Blum M (2003) Neural crest origin of mammalian Merkel cells. Dev Biol 253:258–263

Tachibana T (1995) The Merkel cell: recent findings and unresolved problems. Arch Histol Cytol 58:379–396

Terenghi G, Sundaresan M, Moscoso G, Polak JM (1993) Neuropeptides and a neuronal marker in cutaneous innervation during human foetal development. J Comp Neurol 328:595–603

Vielkind U, Sebzda MK, Gibson IR, Hardy MH (1995) Dynamics of Merkel cell patterns in developing hair follicles in the dorsal skin of mice, demonstrated by a monoclonal antibody to mouse keratin 8. Acta Anat 152:93–109

Merkel Cells Are Postmitotic Cells of Neural Crest Origin

Miloš Grim[1], Zdenek Halata[2], Viktor Szeder[1,3] and Maya Sieber-Blum[3]

[1]Institute of Anatomy, First Medical Faculty, Charles University Prague, Czech Republic
[2]Institute of Anatomy, University of Hamburg, Germany
[3]Department of Cell Biology, Neurobiology and Anatomy, Medical College of Wisconsin, Milwaukee, Wisconsin, USA

Summary

Chick/quail chimeras and double transgenic Wnt1-cre/R26R mice, in which neural crest cells are permanently marked by expression of β-galactosidase, are evidence that Merkel cells in Merkel nerve endings of birds and mammals are neural crest derivatives. Like melanoblasts their precursors colonise the skin. Cytokeratin 8- and 18-positive Merkel cells in mouse whisker follicles are postmitotic cells. They are Ki 67-negative and their nuclei exhibit features of differentiated resting cells. In contrast, Merkel cell progenitors are likely to be proliferating cells. Simple cytokeratins are not suitable markers for their identification. Possible progenitors of Merkel cells that contain dense core granules can be identified using electron microscopy in the connective tissue surrounding the bulge region of the whisker follicle. Possible progenitors of Merkel cells in epidermis are β-galactosidase-positive, but cytokeratin-negative. Verification of the Merkel cell precursor nature of these cells requires specific markers that will enable us to distinguish them from precursors of melanoblasts. Thus, the question of the location of proliferative Merkel cell progenitors remains open.

Introduction

Merkel cells, which are localized in the skin of vertebrates, are large, oval cells with an electron-clear cytoplasm. The vast majority of Merkel cells are in close association with a nerve terminal. Recent experimental evidence points to direct involvement of the Merkel cells in transducing mechanical stimuli to action potentials in the afferent nerve fibre (Senok and Baumann 1997; Tazaki and Suzuki

1998; Halata et al. 2003; see Baumann and Senok, this Vol.).

Merkel cells were originally described by Merkel (1875) in the skin of birds and mammals. The electron-microscopic features of Merkel cells in birds (Saxod 1978; Halata and Grim 1993) are similar to those in mammals (Halata 1970). The main difference between avian and mammalian Merkel cells is their location. In mammals, Merkel cells are found in the basal layer of the epidermis (Munger 1965) and in the outer root sheath of the hair follicle, in particular in whiskers (Andres 1966). By contrast, Merkel cells in birds are localised in the dermal connective tissue (Saxod 1978). However, the epidermis of birds is completely free not only of Merkel cells and their nerve endings, but also of free nerve endings, which terminate in the dermis (Hemming et al. 1994).

There is still controversy about the developmental origin of Merkel cells. According to one view, they originate from the neural crest and migrate into the mammalian epidermis during embryonic development. This opinion is supported by the observation of Merkel cells in the dermis of fetal and newborn mammalian skin (Breathnach and Robins 1970; Hashimoto 1972; Winkelmann 1977). An alternative view is that they arise from an ectodermal stem cell common to keratinocytes and Merkel cells (Munger 1965; English 1974; Moll et al. 1986; Moll and Moll 1992). The latter hypothesis is currently favoured by most investigators. It is mainly supported by the finding of low molecular cytokeratins (CK 8, CK 18, CK 19 and CK 20) in Merkel cells of mammals (Moll et al. 1984, 1995; Kim and Holbrook 1995).

Whisker hair follicles are richly innervated and the nerve fibres end in a variety of different mechanoreceptors, in particular in Merkel nerve endings. Up to 2000 Merkel cells were found in whiskers of the rat, mouse, cat and monkey (Halata and Munger 1980; Halata 1993). In some aquatic mammals, such as in seal and sea lion, their number in a single follicle can reach 20,000 (see Dehnhardt et al., this Vol.). Whiskers are therefore conducive to the study of Merkel cell development.

This study aims at summarizing our recent results on the developmental origin of Merkel cells in embryonic chick/quail chimeras and in mouse whisker follicles and it addresses the question of Merkel cell precursors and their proliferation.

Methods

White Leghorn chicken (*Gallus gallus*) and Japanese quail (*Coturnix coturnix japonica*) were studied and used to prepare embryonic chimeras in order to elucidate the developmental origin of Merkel cells in birds (for details see Grim and Halata 2000). To determine the origin of mammalian Merkel cells and to explore their proliferating precursor population, we have analysed Merkel cells in whiskers of normal and double transgenic Wnt1-cre/R26R mice. In this double transgenic mouse, neural crest cells are permanently marked by β-galactosidase expression (Szeder et al. 2002; see Sieber-Blum et al., this Vol.). The whisker pads of embryonic day (ED) 13.5–18.5 embryos and of new-born mice were dissected, genotyped and double transgenic tissue processed for β-galactosidase immunohistochemistry (Bgal) and/or catalytic Xgal histochemistry. Merkel cells in paraffin and cryostat sections were visualised with antibodies against cytokeratin 8 (CK 8; TROMA-1, Hybridoma Bank) and cytokeratin 18 (CK 18; RGE 53, Chemicon). The cell proliferation marker, Ki 67 protein, was detected using the rat anti-mouse monoclonal antibody, Ki 67 (TEC-3, DAKO, Glostrup, Denmark). This recently developed antibody (Scholzen and Gerdes 2000) can be used on formalin-fixed, paraffin-embedded murine tissue. In addition, ultrastructural features of Merkel cells were studied using electron microscopy (Halata et al. 1999).

Results and Discussion

In our previous study (Grim and Halata 2000), the developmental origin of avian Merkel cells was investigated in chick/quail chimeras at the light microscopy level and by electron microscopy (Fig. 1). Quail cells in chimeras were identified according to the presence of masses of heterochromatin associated with the nucleolus, which is characteristic for quail cells (Le Douarin 1973). Our results show that Merkel cells are neither derived from the ectoderm nor from the mesenchyme of the limb primordium, but that they originate from cells that migrate into the limb from axial structures of the embryo. Two sources of cells are known to colonise the limb: somites and the neural crest. Our results show that Merkel cell progenitors colonised the limb skin simultaneously with glial cells and melanocytes. We thus conclude that avian Merkel cells originate from the neural crest.

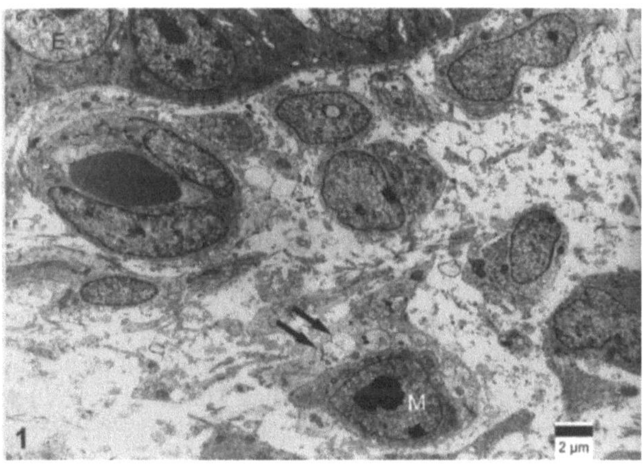

Fig. 1. Merkel cell in the dermis of the foot of a chimeric chick 14 days after grafting a chick limb primordium onto the quail host embryo. Merkel cell (*M*) exhibits heterochromatin mass associated with the nucleolus characteristic for the quail that is not present in surrounding dermal chick cells. The Merkel cell cytoplasm contains a small number of dense core granules. Several axons (*arrows*) are in contact with the cytoplasmic membrane of the Merkel cell. *E* Epidermis of the chick

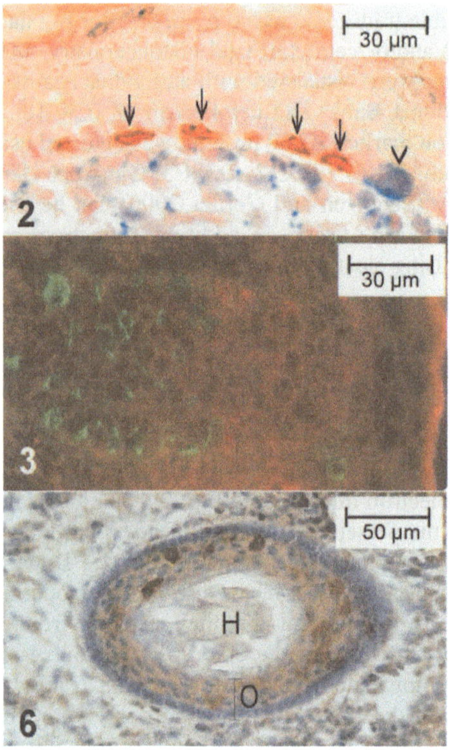

Fig. 2. Cryostat section of the upper lip skin of Wnt1-cre/R262 mouse at postnatal day 4. Cells of neural crest origin express β-galactosidase and are Xgal-positive (*blue* reaction product). Merkel cells in the epidermis express CK 8 (*brownish-red* reaction product). Note co-localisation of Xgal and CK 8 in Merkel cells (*arrows*). Xgal positivity of epidermal cell of basal layer without CK 8 positivity (*arrowhead*)

Fig. 3. Merged image of whisker hair follicle and epidermis of mouse at embryonic day 17. Merkel cells are stained for CK 18 (*fluorescein*) and cell proliferation marker Ki 67 is visualised with rhodamine (*red*). Merkel cells are not double stained

Fig. 6. Whisker follicle of mouse at postnatal day 4 in oblique cryostat section. Cells of neural crest origin expressing β-galactosidase (*brown*) are dispersed throughout the outer root sheath (*O*); *H* hair shaft

Avian Merkel cells thus represent another population of neural crest-derived cells. The developmental origin of mammalian Merkel cells has been determined in the whisker follicle of the double transgenic Wnt1-cre/R26R mouse, in which neural crest cells are permanently marked by expression of β-galactosidase (Szeder et al. 2002; see Sieber-Blum et al., this Vol.). Merkel cells were visualised by staining for CKs 8 and 18, two typical Merkel cell markers (Moll et al. 1984). Our results show that CK 8-positive Merkel cells in mouse whiskers and in the basal layer of the epidermis also express β-galactosidase (Fig. 2). This finding

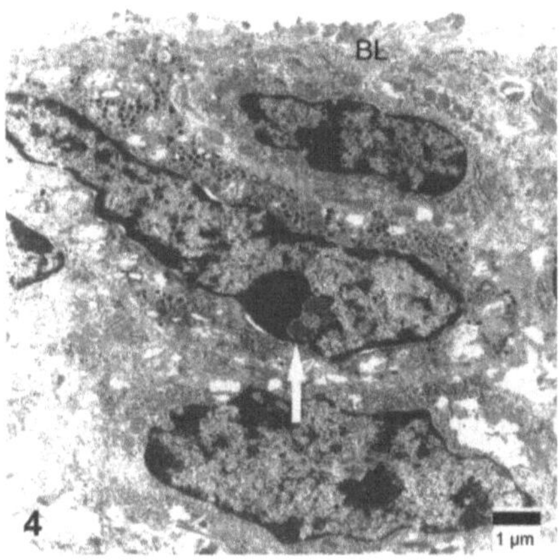

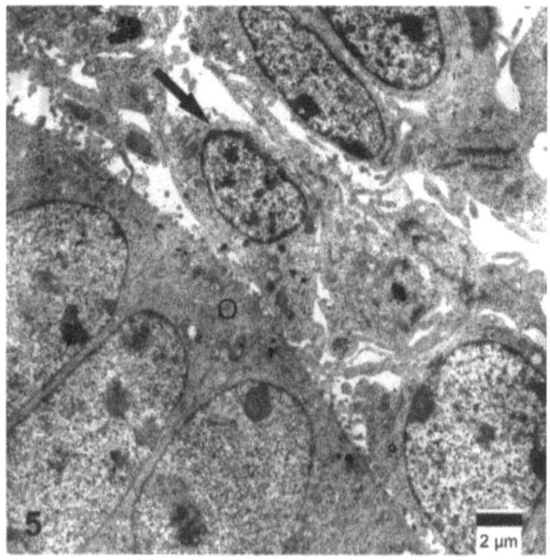

Fig. 4. Merkel cell with numerous dense core granules in the outer root sheath of the mouse whisker at postnatal day 4. Merkel cell exhibits chromatin condensation and the ring-shaped nucleolus (*arrow*) characteristic for mature resting cells. *BL* Basal lamina of hair follicle

Fig. 5. Part of whisker follicle of mouse at embryonic day 16. In connective tissue close to the outer root sheath (*O*), a cell (*arrow*) contains dense core granules of similar size and density as Merkel cells. This cell probably represents a Merkel cell precursor

provides solid evidence for the neural crest origin of mammalian Merkel cells in the whisker follicles of the mouse.

The whisker follicles contain the largest known accumulation of Merkel cells, elements of the mechanosensory Merkel nerve endings (Gottschaldt et al. 1973). The number of Merkel cells in whisker follicles increases during embryonic development (Pasche et al. 1990), but mitotic Merkel cells are observed only rarely. For example, Mérot et al. (1987) studied embryonic mouse whisker follicles by electron microscopy looking for dividing cells identified as Merkel cells by their content of dense core granules. In our current study, we use antibodies against CK 8 and CK 18, markers of Merkel cells, in combination with the Ki67 antibody, a marker for cell proliferation, to determine the cell cycle status of Merkel cells and their progenitors. The Ki 67 epitope is expressed during all active phases of the cell cycle, but is absent in resting cells (Scholzen and Gerdes 2000).

We do not observe any Ki 67 binding to differentiated Merkel cells, suggesting that they are postmitotic (Fig. 3). Our data are in agreement with a report by Moll et al. (1996), who used antibodies against cytokeratins as well as Ki 67 antibody and obtained similar results in the epidermis of human fetal skin. It can thus be concluded that once Merkel cells express simple cytokeratins, they are postmitotic, terminally differentiated cells. Electron microscopic observation of Merkel cell nuclei showing ring-shaped nucleoli and chromatin condensation (Fig. 4) supports this notion. Ring-shaped nucleoli are characteristic for resting cells (Smetana 1974).

In a subsequent series of experiments, we were searching by electron microscopy for cells that might represent Merkel cell progenitors. Ultrastructurally, Merkel cells can be identified by their content of dense core granules 60–80 nm in diameter and by their association with nerve terminals. In this study we find cells in the dermis containing a low number of dense core granules corresponding in size to the granules of mature Merkel cells (Fig. 5). They are localized in connective tissue surrounding the whisker follicle in close proximity to its bulge region. They are not in contact with nerve fibres. However, in cryostat sections stained for CKs 8 and 18, we have never observed cytokeratin-positive cells in the dermis. Similar results were reported by Pasche et al. (1990). Cells with dense core granules, interpreted as representing Merkel cells, were observed in dermis of human embryonic skin (Breathnach and Robins 1970; Hashimoto 1972) and newborn cats (Halata 1981). In Merkel cells of the mouse, dense core granules appear earlier than staining for simple cytokeratins is detectable. For example, Pasche et al. (1990) have observed dense core granules in Merkel cells in ED 12 whisker follicles, but CK 18 immunoreactivity was first detectable at ED 13 only.

Other cells that might represent Merkel cell progenitors are observed within the basal layer of the epidermis and in the outer root sheath of Wnt1-cre/R26R double transgenic whisker follicles. They express the neural crest cell marker, the β-galactosidase, but are negative for CK 8 staining (Fig. 2). Moreover, β-galactosidase expressing neural crest-derived cells is dispersed throughout the entire width of the epithelium of the multilayered outer root sheath in the bulge region (Fig. 6). The majority of these cells is not localised in the outermost layer of the outer root sheet, where the cytokeratin expressing Merkel cells are typically localised. These cells may represent Merkel cell precursors or melanogenic progenitors, two neural crest cell types that colonise the epidermis. Melanocytes can be identified by the presence of melanin granules within the cytoplasm. However, in the cytoplasm of Bgal-positive and CK-negative cells, no melanin granules are present at this developmental stage. Only the availability of a marker specific for the Merkel cell lineage will permit the study of the proliferative behaviour of Merkel cell progenitors.

The question where Merkel cell progenitors are located when they are still proliferative, remains to be answered. They cannot be identified using simple cytokeratins as markers and other markers specific for the Merkel cell lineage are currently not available. The transcription factors Math1 (Helms et al. 2000; Leonard et al. 2002) and Brn3c (Leonard et al. 2002), which are expressed in mature Merkel cells may provide a future means for the identification of Merkel cell precursors.

Acknowledgements

We thank Eva Kluzáková, Brigitte Asmus and Joan Ward for excellent technical assistance and Dr. O. Nanka for excellent technical assistance in preparing the figures for this manuscript. We thank A.P. McMahon, P. Soriano and H. Sucov for providing the Wnt1-cre and R26R mouse lines. Supported by USPHS grant NS38500 (MSB) and postdoctoral fellowship 1F05-11111-02 (VS) from the National Institute of Neurological Disorders and Stroke, NIH, and by grants VZ 1111 00003-3G (MG) and LN 00A065 (MG) from the Ministry of Education of the Czech Republic and by DAAD (ZH).

References

Andres KH (1966) Über die Feinstruktur der Rezeptoren an Sinushaaren. Z Zellforsch 75:339–365

Breathnach AS, Robins J (1970) Ultrastructural observation on Merkel cells in human fetal skin. J Anat 106:411

English KB (1974) Cell types in cutaneous type I mechanoreceptors (Haarscheiben) and their alterations with injury. Am J Anat 141:105–126

Gottschaldt K-M, Iggo A, Young DW (1973) Functional characteristics of mechanoreceptors in sinus hair follicles of the cat. J Physiol 235:287–315

Grim M, Halata Z (2000) Developmental origin of avian Merkel cells. Anat Embryol 202:401–410

Halata Z (1970) Zu den Nervenendigungen (Merkelsche Endigungen) in der haarlosen Nasenhaut der Katze. Z Zellforsch 106:51–60

Halata Z (1981) Postnatale Entwicklung sensibler Nervenendigungen in der unbehaarten Nasenhaut der Katze. Bibl Anat 19:210–235

Halata Z (1993) Die Sinnesorgane der Haut und der Tiefensensibilität. In: Niethammer J, Schliemann H, Starck D, Wermuth H (eds) Handbook of zoology – Eine Naturgeschichte der Stämme des Tierreiches. De Gruyter, Berlin, pp 1–81

Halata Z, Grim M (1993) Sensory nerve endings in the beak skin of Japanese quail. Anat Embryol 187:131–138

Halata Z, Munger BL (1980) The ultrastructure of the Ruffini and Herbst corpuscles in the articular capsule of domestic pigeon. Anat Rec 198:681–692

Halata Z, Grim M, Christ B (1990) Origin of spinal cord meninges, sheaths of peripheral nerves, and cutaneous receptors including Merkel cells. An experimental and ultrastructural study with avian chimeras. Anat Embryol 182:529–537

Halata Z, Cooper BY, Baumann KI, Schwegmann C, Friedman RM (1999) Sensory nerve endings in the hard palate and papilla incisiva of the goat. Exp Brain Res 129:218–228

Halata Z, Gim M, Baumann KI (2003) Friedrich Sigmund Merkel and his "Merkel Cell", morphology, development and physiology – review and new results. Anat Rec 271:225–239

Hashimoto K (1972) The ultrastructure of the skin of human embryos. X. Merkel tactile cells in the finger and nail. J Anat 111:99–120

Helms AW, Abney AL, Ben-Arie N, Zoghbi HY, Johnson JE (2000) Autoregulation and multiple enhancers control *Math1* expression in the developing nervous system. Development 127:1185–1196

Hemming FJ, Pays L, Soubeyran A, Larruat C, Saxod R (1994) Development of sensory innervation in chick skin: comparison of nerve fibre and chondroitin sulphate distributions in vivo and in vitro. Cell Tissue Res 277:519–529

Kim DK, Holbrook KA (1995) The appearance, density, and distribution of Merkel cells in human embryonic and fetal skin: their relation to sweat gland and hair follicle development. J Invest Dermatol 104:411–416

Le Douarin NM (1973) A Feulgen-positive nucleolus. Exp Cell Res 77:459–468

Leonard JH, Cook AL, Van Gele M, Speleman F, Sturm RA (2002) Expression of developmentally regulated factors in Merkel cell carcinoma. Int J Cancer 101:103–110

Merkel F (1875) Tastzellen and Tastkoerperchen bei den Hausthieren und beim Menschen. Arch Mikr Anat 11:636–652

Mérot Y, Carraux P, Saurar JH (1987) Merkel cell mitosis in vibrissae: an ultrastructural study. J Anat 153:241–244

Moll I, Moll R (1992) Early development of human Merkel cells. Exp Dermatol 1:180–184

Moll I, Moll R, Franke WW (1986) Formation of epidermal and dermal Merkel cells during human fetal skin development. J Invest Dermatol 87:779–787

Moll I, Kuhn C, Moll R (1995) Cytokeratin 20 is a general marker of cutaneous Merkel cells while certain neuronal proteins are absent. J Invest Dermatol 104:910–915

Moll I, Zieger W, Schmelz M (1996) Proliferative Merkel cells were not detected in human skin. Arch Dermatol Res 288:184–187

Moll R, Moll I, Franke WW (1984) Identification of Merkel cells in human skin by specific cytokeratin antibodies: changes of cell density and distribution in fetal and adult plantar epidermis. Differentiation 28:136–154

Munger BL (1965) The intraepidermal innervation of the snout skin of the opossum. J Cell Biol 26:79–97

Pasche F, Mérot Y, Carraux P, Saurat JH (1990) Relationship between Merkel cells and nerve endings during embryogenesis in the mouse epidermis. J Invest Dermatol 95:247–251

Saxod R (1978) Development of cutaneous sensory receptors in birds. In: Jacobson M (ed) Handbook of sensory physiology. Development of sensory systems. Springer, Berlin Heidelberg New York, pp 337–417

Scholzen T, Gerdes J (2000) The Ki-67 protein: from the known and the unknown. J Cell Physiol 182:231–322

Senok SS, Baumann KI (1997) Functional evidence for calcium-induced calcium release in isolated rat vibrissal Merkel cell mechanoreceptors. J Physiol 500:29–37

Smetana K (1974) The ultrastructural morphology and formation of nuclear ribonucleoprotein structures. Acta Fac Med Univ Purk Brunenesis 49:155–179

Szeder V, Grim M, Halata Z, Sieber-Blum M (2002) Neural crest origin of mammalian Merkel cells. Dev Biol 253:258–263

Tazaki M, Suzuki T (1998) Calcium inflow of hamster Merkel cells in response to hyposmotic stimulation indicate a stretch activated ion channel. Neurosci Lett 243:69–72

Winkelmann RK (1977) The Merkel cell system and comparison between it and the neurosecretory or APUD cell system. J Invest Dermatol 69:41–46

Dissociation, Quantification and Culture of Normal Human Merkel Cells Among Epidermal Cell Populations Derived from Glabrous and Hairy Skin Sites

Julie Fradette[1], Danielle Larouche[1], Claudia Fugère[1], Rina Guignard[1], Annie Beauparlant[1], Véronique Couture[1], Louise Caouette-Laberge[2], Alphonse Roy[1] and Lucie Germain[1]

[1]Laboratoire de Recherche des Grands Brûlés/LOEX and Service de Chirurgie, Hôpital du Saint-Sacrement du CHA, Québec, Québec, Canada, G1S 4L8 and Département de Chirurgie, Université Laval, Sainte-Foy, Québec, Canada
[2]Service de Chirurgie, Hôpital Sainte-Justine, Montréal, Québec, Canada

Summary

Merkel cells constitute a unique population that remains difficult to characterize in human skin because of their scarcity. Our aim was to develop tools for the study of Merkel cells in vitro. As a first step, we evaluated the possibility of harvesting human Merkel cells with the two-step extraction method that is widely used to extract and culture keratinocytes. Merkel cells were identified in the epithelial portion of hairy or glabrous skin biopsies by keratin (K)18 and K20 labeling. The totality of cutaneous epithelial cells were isolated from either hairy or glabrous skin biopsies following enzymatic dissociation of both the epidermis and the hair follicles. Flow cytometry was performed to quantify the small Merkel cell population. The analysis revealed that K18-labeled cells represented between 4.0 and 7.6% of freshly dissociated basal epidermal cells. No significant differences were seen between samples derived from glabrous palmar and hairy anatomic sites from children and adults, respectively. We also reported on the presence of Merkel cells in primary and first subcultures of human epidermal cells. The next step will be to enrich the isolated human Merkel cells and improve their culture conditions. An amplification of the number of Merkel cells will allow further studies to unravel long-standing questions regarding their origin, proliferative capacity, and functions in cutaneous biology.

Introduction

The Merkel cell population possesses many interesting features such as coexpression of dense-core neurosecretory granules, keratin (K) filaments and desmosomal proteins. In post-natal skin, Merkel cells express the simple epithelial keratins K18 and K20, and antibodies specific to these keratins enabled their localization in situ (Moll et al. 1984, 1995; Moll 1994; Fradette et al. 1995). Therefore, the distribution and three-dimensional organization of Merkel cells in epidermis and hair follicles have been well described (Lacour et al. 1991; McKenna Boot et al. 1992; Moll 1994; Narisawa et al. 1994). However, the quantification of such a minor cell population remains a difficult task.

We took advantage of our isolation procedure that dissociates cells from both the epidermis and hair follicles (Germain et al. 1997) to quantify Merkel cells in populations of epidermal cells freshly dissociated from human skin. Our interest in Merkel cells is related to the potential improvements in the production and quality of the epidermal grafts produced for the treatment of patients with extensive burns. Moreover, such populations of dissociated cells would further facilitate the study of Merkel cells in relation to stem cells using in vitro models. In this paper, we compared the percentage of Merkel cells (expressing K18) present in freshly dissociated populations of epidermal cells from glabrous (palmar/plantar) and hairy (breast, scalp) anatomic sites using flow cytometry. The presence of dissociated normal human Merkel cells in epidermal cell populations suitable for culture will allow further investigation of their functions, proliferative capacity and fate in vitro.

Results and Discussion

Merkel Cells Are Extracted from Epidermis and Hair Follicles Using a Two-Step Epidermal Cell Isolation Method

When frozen sections of normal human digital skin are labeled with antibodies directed against K20 (Fig. 1A) or K18 (Fig. 1D), positive cells are found in the basal layer of the epidermis, more specifically at the bottom of the rete ridges (Moll et al. 1984; Fradette et al. 1995). The basal layer of interfollicular epidermis as well as the outer root sheath (ORS) of hair follicles also contain numerous Merkel cells (Fig. 1J; Lacour et al. 1991; Narisawa et al. 1993, 1994; Moll 1994; Fradette et al. 1995). Since Merkel cells are often found in close association with dermal nerve endings, we first verified if a commonly used two-step epidermal cell extraction method allowed the recovery of viable Merkel cells. This method was originally designed for the isolation of keratinocytes, free of fibroblasts, and allows massive cell amplification for the production of cultured epithelial sheets suitable for grafting (Green et al. 1979; Germain et al. 1993).

We first determined whether Merkel cells remained in the epithelial portion of skin after thermolysin treatment and dermo-epidermal separation using immunohistochemistry on frozen sections. Figure 1G demonstrates that K20-expressing

Merkel cells are indeed retained in the basal layer of the epidermal portion of palmar skin. Similarly, transversal sections at the level of the upper follicle (Fig. 1K) clearly identify K20-expressing Merkel cells in the basal layer of the ORS after scalp skin separation following thermolysin incubation.

Quantification of Merkel Cells from Glabrous and Hairy Anatomic Sites

To quantify Merkel cells, the percentage of K18-expressing cells among the epidermal population freshly dissociated from glabrous (palmal/plantar) or hairy (breast, scalp) skin biopsies was determined. Briefly, skin biopsies (3–9 cm^2) were cut into 5×5 mm fragments and incubated overnight at 4 °C or 3 h at 37 °C in a 500 μg/ml thermolysin solution (Germain et al. 1993). The epithelium (epidermis and hair follicles) was separated from the underlying dermis with fine forceps and the epidermal cells were dissociated with trypsin for 15 min at 37 °C. The freshly isolated epidermal cell suspensions were fixed with 70% cold ethanol before immunolabeling. Flow cytometry analysis confirmed that Merkel cells represent a minor population of the total epidermal cells, with 1.6±0.5% of K18-expressing cells for samples isolated from glabrous skin of children and 2.0±0.2% for hairy skin biopsies from adults. To ensure adequate comparisons between the dissociated cell populations obtained from different body sites, the percentage of basal cells was determined for each sample by labeling with the VM-2 antibody (VM2, clone #HB-8530 ATCC, Manassas, VA) which recognizes the α3-integrin subunit expressed by basal cells of the epidermis and ORS of hair follicles (Kaufmann et al. 1989; Michel et al. 1996).

Table I. Percentage of human Merkel cells in biopsies from different anatomic sites

Anatomic site (skin)	Age of donor (Years) (Mean ± SD)	Percentage of epidermal basal cells expressing K18* (% ± SD)
Palmar/plantar (n=7)	1.6 ± 0.5	4.0 ± 0.6
Breast (n=4)	40.6 ± 12.4	4.8 ± 2.1
Scalp (n=4)	46.5 ± 8.2	7.6 ± 3.2

* The percentage of basal cells that are Merkel cells was evaluated from VM2 and K18 labelings. Results are expressed as mean ± SD of (%K18 positive cells/ %VM2 positive cells) x 100. Cells were analyzed on a Becton-Dickinson fluorescence-activated cell sorter (FACS) and a number of 20 000 events was acquired for each sample.

Table 1 represents the percentage of freshly dissociated basal cells that express K18. The Merkel cell population, as identified by K18 expression, corresponds to 4.0% of basal cells for palmar skin of children and is not statistically different from the percentages obtained for hairy skin biopsies from adults (Table 1, $p<0.05$ according to nonparametric Mann-Whitney test). This is interesting since glabrous

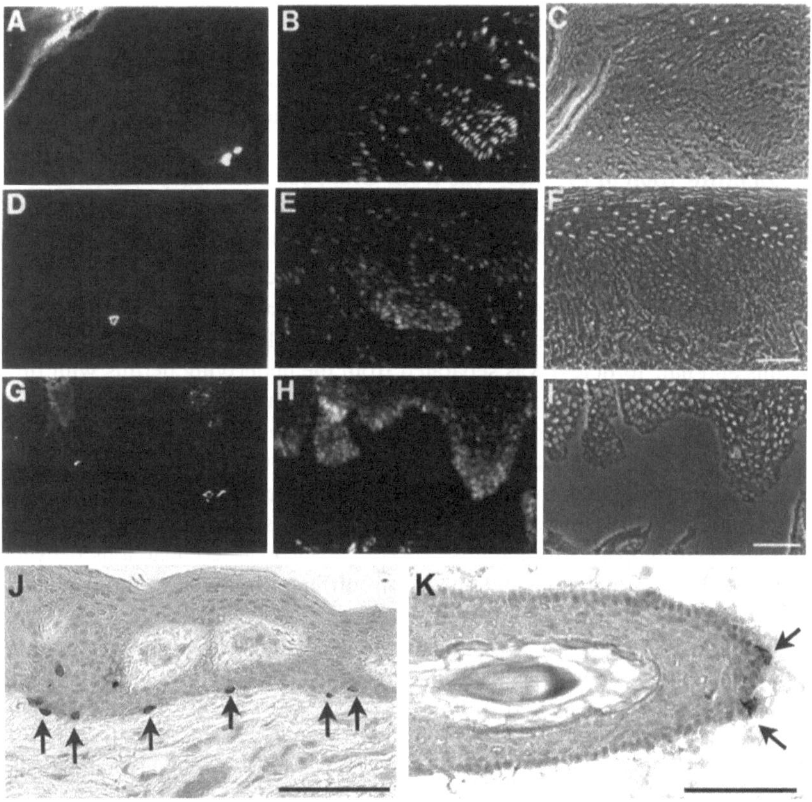

Fig. 1A–K. Merkel cells in human palmar epidermis and hair follicles are conserved in the basal layer after thermolysin treatment. Indirect immunofluorescence labeling of 6-μm acetone-fixed frozen sections using antibodies directed against **A**, **G** K20 (IT-Ks 20.10, American Research Product, Belmont, MA) and **D** K18 (K18.174, Progen-IBL, Cambridge, MA) allowed detection of Merkel cells. Merkel cells are localized within the basal layer of the normal palmar epidermis **A** before and **G** after cleavage of the dermo-epidermal junction by thermolysin. **B**, **E**, **H** Nuclear Hoechst staining (reagent 33258, Sigma); **C**, **F**, **I** phase-contrast micrographs corresponding to **A**, **D**, **G**, respectively. Similarly, thermolysin treatment of hair follicles containing biopsies such as adult scalp skin allows the separation of intact hair follicles from the dermal compartment and isolation of follicular Merkel cells: indirect peroxidase labeling was performed on formaldehyde (2.5%, 30 min at RT) and methanol (10 min at –20 °C) fixed frozen sections using the Ultra HRP detection system (ID Labs Inc, London, Ontario, Canada). K20-expressing cells are found in the basal layer of the epidermis and in the outer root sheath of terminal scalp hair follicle **J** before, and **K** after thermolysin separation. *Scale bars*: **A–F** 47 μm; **G–I** 62 μm; **J–K** 100 μm

skin was considered to have a greater density of Merkel cells per mm^2 or cm^2 of basal layer (Lacour et al. 1991; McKenna Boot et al. 1992). The high number of

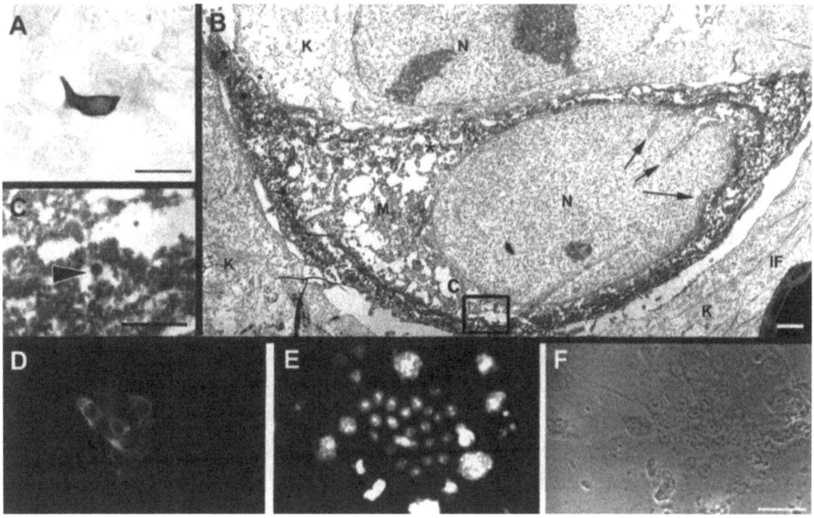

Fig. 2A–F. Merkel cells are present in culture in vitro. **A–C** Detection of K20-expressing cells in cultures in vitro showing that Merkel cells are with keratinocyte colonies. Follicular Merkel cells are extracted using the two-step epidermal cell isolation procedure and culture in vitro as described in Fradette et al. (2003). **A** Detection of Merkel cell expressing K20 in epithelial cell culture derived from human scalp skin. The electron microscopy image (**B**) represents the labeled cell on the light microscopy image in **A**. **C** Magnification of the delimited region *c* of **B**. The cytoplasmic dense peroxidase deposits (*asterix*) allowed the identification of a K20-expressing cell at the electron microscopy level (**B**). In contrast, large keratinocytes presented a light cytoplasm containing an extensive network of keratin filaments (**B**). The presence of discrete dense-cored neurosecretory granules (*arrowheads*, **C**) in the cytoplasm of the K20-labeled cell indicates that they are Merkel cells. Moreover, the characteristic lobulated nucleus of Merkel cells is indicated by *arrows* (**B**). **D–F** In vitro, K18 is induced in keratinocytes cultured as monolayers. **D** Primary cultures (day 5) of epidermal cells derived from plantar skin were stained with an antibody directed against K18. Human cells forming the central colony are distinguished from the murine 3T3 feeder layer by the uniform appearance of their nuclei revealed by Hoechst staining (**E**). **F** The corresponding phase-contrast micrograph. Epidermal cells were plated on a mouse irradiated 3T3 fibroblast feeder layer in culture flasks (27,000 cells/cm^2) or coverslips (80,000 cells/cm^2) and cultured in complete DME-HAM supplemented with 10% Fetal Clone II serum (Hyclone, PDI Bioscience, Aurora, Ontario, Canada; Rheinwald and Green 1975; Germain et al. 1993; Michel et al. 1996). *IF* Intermediate filaments; *K* keratinocyte; *M* Merkel cell; *N* nucleus. *Scale bars*: **A** 50 µm; **B** 1 µm; **C** 500 nm; **D–F** 47 µm. Part of this figure is from Fradette et al. (2003) J Invest Dermatol, Courtesy of Blackwell Publishing

Merkel cells we obtained from adult hairy skin sites is likely due to the contribution of Merkel cells present in many small vellus hair follicles (breast) and terminal hairs (scalp) which are easily extracted by our technique (Fig. 1; Germain et al. 1997). Since skin presents a complex architecture in which Merkel cells are not distributed regularly, it is difficult to obtain representative quantification from tissue sections. In contrast, flow cytometry offers the advantage of analysing a large number of cells. Therefore, we determined the percentage of Merkel cells among epidermal cells freshly dissociated from a relatively large representative skin area (up to 9 cm^2) containing numerous hair follicles when applicable.

Primary cultures from each anatomic site were established by seeding dissociated epidermal cells with a 3T3 feeder layer, either at high density on glass coverslips for immunolabeling, or into flasks for amplification and analysis by flow cytometry. Numerous positive cells were observed when the monolayer cultures were immunostained with an anti-K18 antibody, likely representing induction of K18 expression in keratinocytes under our culture conditions (Fig. 2D–F). The extent of this reexpression was variable between anatomic sites and was particularly important for children finger skin where up to 77% of keratinocytes express K18 at the end of primary culture (Table 2). Such reexpression of K18 has also been observed for cultured epithelial sheets (Compton et al. 1990), reinforcing the notion that K18 is not a reliable marker for the identification of Merkel cells in cultures since it has lost the specificity for Merkel cells observed in situ, as well as in cells freshly dissociated from children and adult skin biopsies. Thus, K20 is a more reliable marker for the identification of Merkel cells in culture, whereas K18 remains an appropriate marker for the identification of Merkel cells in situ and among freshly dissociated epidermal cells before culture. We have been able to detect human Merkel cells within cultures of epidermal cells in vitro by K20 immunostaining (Fig. 2A–C). The electron microscopy analysis confirmed the ultrastructural features of Merkel cells: lobulated nucleus, dense-cored granules and cytoplasmic extensions (Fig. 2B,C; Fradette et al. 2002).

The optimization of culture conditions for human Merkel cells will likely provide a unique in vitro model to gain insight into their functions and controversial features. Moreover, the presence of Merkel cells in culture could be beneficial for patients since a high number of Merkel cells in cultured epithelial sheets grafted to extensively burned patients could stimulate reinnervation and help them regain better sensory functions at targeted grafted sites. With the rapid progress made in the fields of tissue engineering (Auger et al. 2000), transplantation and gene therapy, tissue reconstruction using the various cell types present in skin holds great promise and could result in the production of skin substitutes tailored for particular needs and specialized applications.

Table II. Percentage of K18-expressing cells in monolayer cultures of epidermal cells derived from various anatomic sites

Anatomic site (skin)	Palmar/Plantar	Breast	Scalp
Primary culture (% ± SD)	77.4 ± 8.5	2.6 ± 1.6	25.1 ± 9.7
Passage 1 (% ± SD)	87.7 ± 4.9	5.9 ± 0.5	45.8 ± 12.2

Acknowledgements

This work was supported by the Canadian Institute of Health Research Grant MT-12087 to L.G. J. F. and D.L. were recipient of studentships from the Medical Research Council of Canada. L.G. is the holder of a Canadian Research Chair on stem cells and tissue engineering.

References

Auger FA, Remy-Zolghadri M, Grenier G, Germain L (2000) The self-assembly approach for organ reconstruction by tissue engineering. Biomed J Regenerative Med 1:75–86

Compton CC, Regauer S, Seiler GR, Landry DB (1990) Human Merkel cell regeneration in skin derived from cultured keratinocyte grafts. Lab Invest 62:233–2241

Fradette J, Godbout M-J, Michel M, Germain L (1995) Localization of Merkel cells at hairless and hairy human skin sites using keratin 18. Biochem Cell Biol 73:635–639

Fradette J, Larouche D, Fugère C, Guignard R, Beauparlant A, Couture V, Caouette-Laberge L, Roy A, Germain L (2003) Normal human Merkel cells are present in epidermal cell populations isolated and cultured from glabrous and hairy skin sites. J Invest Dermatol 120(2):313–317

Germain L, Rouabhia M, Guignard R, Carrier L, Bouvard V, Auger FA (1993) Improvement of human keratinocyte isolation and culture using thermolysin. Burns 2:99–104

Germain L, Michel M, Fradette J, Xu W, Godbout M-J, Li H (1997) Skin stem cell identification and culture: a potential tool for rapid epidermal sheet production and grafting. In: Rouabhia M (ed) Skin substitute production by tissue engineering: clinical and fundamental applications. Landes Bioscience, Austin, TX, pp 177–210

Green H, Kehinde O, Thomas J (1979) Growth of cultured human epidermal cells into a multiple epithelia suitable for grafting. Proc Natl Acad Sci USA 76:5665–5668

Kaufmann R, Frosch D, Westphal C, Weber L, Klein C (1989) Integrin VLA-3: ultrastructural localization at cell-cell contact sites of human cell cultures. J Cell Biol 109:1807–1815

Lacour JP, Dubois D, Pisani A, Ortonne JP (1991) Anatomical mapping of Merkel cells in normal human adult epidermis. Br J Dermatol 125:535–542

McKenna Boot P, Rowden G, Walsh N (1992) The distribution of Merkel cells in human fetal and adult skin. Am J Dermatopathol 14:391–396

Michel M, Török N, Godbout M-J, Lussier M, Gaudreau P, Royal A, Germain L (1996) Keratin 19 as a biochemical marker of skin stem cells in vivo and in vitro: keratin 19 expressing cells are differentially localized in function of anatomic sites, and their number varies with donor age and culture stage. J Cell Sci 109:1017–1028

Moll I (1994) Merkel cell distribution in human hair follicles of the fetal and adult scalp. Cell Tissue Res 277:131–138

Moll I, Kuhn C, Moll R (1995) Cytokeratin 20 is a general marker of cutaneous Merkel cells while certain neuronal proteins are absent. J Invest Dermatol 104:910–915

Moll R, Moll I, Franke WW (1984) Identification of Merkel cells in human skin by specific cytokeratin antibodies: changes of cell density and distribution in fetal and adult plantar epidermis. Differentiation 28:136–154

Narisawa Y, Hashimoto K, Nakamura Y, Kohda H (1993) A high concentration of Merkel cells in the bulge prior to the attachment of the arrector pili muscle and the formation of the perifollicular nerve plexus in human fetal skin. Arch Dermatol Res 285:261 268

Narisawa Y, Hashimoto K, Kohda H (1994) Merkel cells of the terminal hair follicle of the adult human scalp. J Invest Dermatol 102:506–510

Rheinwald JG, Green H (1975) Serial cultivation of strains of human epidermal keratinocytes: the formation of keratinizing colonies from single cells. Cell 6:331–343

Growth of Nerve Fibers to Merkel Cells Observed in Co-Culture of Sensory Ganglia and Sinus Hair Follicles

Jun Fukuda, Hisako Ishimine, Kazuko Keino-Masu, Yoshiaki Masaki

Laboratory of Molecular and Cellular Physiology, Department of Physiology, National Defense Medical College of Japan, 3–2 Namiki, Tokorozawa, Saitama 359-8513, Japan

Summary

Merkel cells have long been assumed to exhibit a function of guiding sensory nerve fibers to the skin and hairs. However, there has been little experimental evidence that supports this hypothesis. In order to test this possibility, we performed co-culture of sinus hair follicles and sensory ganglia in a collagen gel, intending to examine how sensory nerve fibers grow to the Merkel cell-enriched regions of sinus hair follicles. During co-culture with a serum-free medium supplemented with NGF and NT3, nerve fibers from sensory ganglia grew in a manner making convergence to the superior enlargement of hair follicles at which Merkel cells were enriched. Live Merkel cells stained with FM1-43 dye were demonstrated to be present in co-culture preparations. The presence of Merkel cells in this region of hair follicles was also demonstrated by immunostaining against cytokeratin 20. Immunostaining of a co-culture preparation against synaptophysin 1 showed that the hair follicle was surrounded tightly with synaptophysin 1-enriched newly grown sensory nerve fibers. We therefore considered that Merkel cells are likely to have a capability of attracting sensory nerve fibers.

Introduction

Merkel cells are innervated by primary sensory nerve fibers generating from sensory ganglia of trigeminal and dorsal spinal roots (Andres 1966). It has long been assumed that one of the functions of Merkel cells is to guide the sensory nerve fibers to the skin and hairs (Lumsden and Davies 1986). Vos et al. (1991) reported that Merkel cells in monolayer culture tend to form a tight contact with sensory nerve fibers. However, no further experimental evidence has been provided whether Merkel cells are capable of attracting sensory nerve fibers. In order

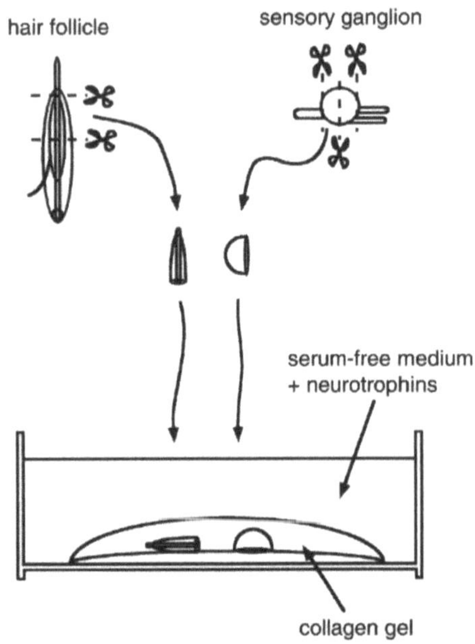

Fig. 1. Methods of co-culture of hair follicles and sensory ganglia in collagen gel. Shafts were dissected from sinus hair follicles by removing their capsules. Dorsal root ganglia were dissected and cut into halves. These follicular and DRG preparations were embedded in collagen gel and were cultured with a serum-free medium supplemented with neurotrophins, NGF, BDNF, NT3 or NT4 in a CO_2 incubator

to provide an answer to this question, we planned a co-culture experiment of sensory ganglia and epidermal preparations that contained a large number of Merkel cells. In this experiment we assumed that if Merkel cells exhibited the capability of attracting sensory nerves, fibers grown from the sensory ganglia tended to grow toward the epidermal preparations. Indeed, in the co-culture, sensory nerve fibers that extended from the ganglia turned their growing direction and then converged onto several regions on the epidermal preparations at which Merkel cells were enriched. A series of experimental results represented in this symposium provide evidence of the capability of Merkel cells of attracting sensory nerve fibers.

Methods

Co-Culture of Hair Follicles and Sensory Ganglia in Collagen Gel

Co-culture of sinus hair follicles and sensory ganglia was performed as shown schematically in Fig. 1. Sinus hair follicles were dissected from whisker pads of newborn and infant rats. Shafts that contain a large number of Merkel cells were dissected from the hair follicles by removing their capsules. Sensory ganglia, either dorsal root ganglia (DRG) or trigeminal ganglia (TRG), were dissected from another rats. Connective tissues surrounding the ganglia were removed. These ganglia were cut into halves to obtain ganglia of half-domed shape. The hair follicle and ganglion preparations were embedded in collagen gel and were incubated with a serum-free medium supplemented with neurotrophins, either NGF (20 ng/ml), BDNF (20 ng/ml), NT3 (20 ng/ml) and NT4 (20 ng/ml). The culture medium contained DMEM/F-12 Medium, 12 mM $NaHCO_3$, 0.1 mg/ml transferrin, 99.3 μg/ml putrescine, 0.87 μM insulin, 0.02 μM progesterone and 1% penicillin-streptomycin (Fukuda 1996).

Immunostaining of Co-Culture Preparations

Co-culture preparations were fixed with 4% paraformaldehyde and were immunostained against 160-kDa neurofilament, cytokeratin 20 or synaptophysin 1. The immunostained preparations were colored by DAB. Some preparations were examined under a fluorescence microscope, in which Cy3-conjugated secondary antibodies were used.

Identification of Live Merkel Cells in Culture by Means of Staining with FM1-43 dye

The methods of staining live Merkel cells with FM1-43 dye are described in an accompanying paper of this symposium (see Fukuda et al., this Vol.).

Results

Growth of Sensory Nerve Fibers to Sinus Hair Follicles in Co-Cultures

Figure 2 demonstrates a result of co-culture of a DRG and two hair follicles in collagen gel with serum-free medium supplemented with NGF (20 ng/ml) and NT3 (20 ng/ml) for 3 days. In order to visualize nerve fibers growing in the collagen gel, the co-culture preparation was immunostained against 160-kDa neurofilament and colored by DAB. It could be seen that nerve fibers growing from the cut ends of distal and proximal nerves of the DRG change direction (open arrows) and then grow toward the follicle preparations. The result of this co-culture indicates that sensory nerve fibers regenerating from the DRG appear to be attracted by the follicular preparations.

Nerve fibers from the trigeminal ganglion (TRG) also represent a growth pattern as if they are attracted by hair follicles (Fig. 3). Many nerve fibers from the TRG (from P3 rat) are seen to grow in a straight direction to a hair follicle of the same animal during co-culture for 3 days. A nerve fiber indicated by a filled arrow shows an apparent change of direction growing toward the hair follicle. On the surface of the hair follicle, a newly formed network of nerve fibers during the co-culture is seen. Many nerve fibers appear to grow circling the surface of the hair follicle while some leave the hair follicle in a straight manner (filled arrowheads).

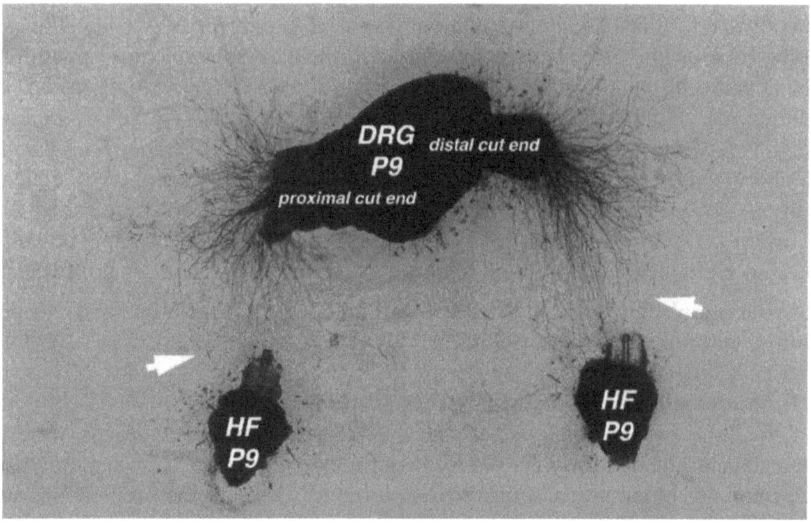

Fig. 2. Growth of DRG fibers to superior enlargement of sinus hair follicles. A *DRG* and two hair follicles (*HF*) dissected from a P9 rat were co-cultured in collagen gel with NGF (20 ng/ml) and NT3 (20 ng/ml) for 3 days. The preparation was immunostained against a 160-kDa neurofilament and colored by DAB. Nerve fibers growing from the cut ends of both the proximal and distal nerves are seen to converge on the HF preparations grown in the collagen gel, the co-culture preparation was immunostained against a 160-kDa neurofilament and colored by DAB. It can be seen that nerve fibers growing from the cut ends of distal and proximal nerves of the DRG change direction (*open arrows*) and grow toward the follicle preparations. The result of this co-culture indicates that sensory nerve fibers regenerating from the DRG appear to be attracted by the follicular preparations

In order to examine whether synapse-like structures were formed in co-culture, a co-culture preparation was immunostained against synaptophysin 1 that is known to be a protein located specifically in presynaptic nerve terminals. When an immunostained preparation in which the secondary antibody was conjugated with Cy3 was examined under a confocal laser microscope (Fig. 4), synaptophysin 1 was seen enriched on the surface of a hair follicle (Fig. 4A, arrowheads). In an image of its cross section constructed by a computer calculation (Fig. 4B), synaptophsyin 1 (Ryan 2001) is seen condensed (arrowhead) on the surface of the hair follicle. These observations indicate that synaptic-like structures are likely to be formed on the hair follicle in co-culture.

Identification of Merkel Cells in Hair Follicles

Merkel cells are considered to be one of the candidate cells that cause attraction of sensory nerve fibers to hair follicles in co-culture. We thus examined whether Merkel cells were indeed present in hair follicles to which sensory nerve fibers converged. As we have demonstrated in a paper presented in this symposium (see Fukuda et al., this Vol.), live Merkel cells are capable of incorporating FM1-43

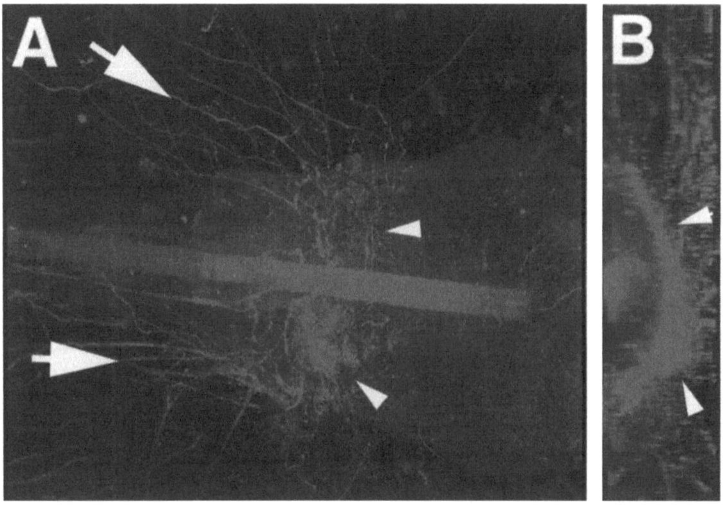

Fig. 3. Newly formed innervation of sensory fibers to a sinus hair follicle. A shaft of a hair follicle from a P3 rat was co-cultured with a half-dome-shaped trigeminal ganglion (*TRG*) of the same animal in collagen gel for 3 days. Many nerve fibers (*open arrows*) can be seen to converge on the superior enlargement of the follicle preparation. Some nerve fibers (*filled arrow*) can be seen to turn their growing direction toward the hair follicle

that has been well known to be a fluorescent dye incorporated into presynaptic nerve terminals during recycling of synaptic vesicles (Sudhof 2000). When a hair follicle was stained with FM1-43 dye, spindle-like cells representing yellow-green fluorescence (Fig. 5A) under a fluorescent microscope are seen lined up on the surface of a hair follicle.

In another hair follicle (Fig. 5B) whose Merkel cells were immunostained against cytokeratin 20 (Tachibana et al. 2000) and colored by Cy3, red-colored spindle-like cells were seen lined up in the hair follicle under a fluorescent microscope. Since cytokeratin 20 is known to be a protein specifically expressed in Merkel cells, they were concluded to be Merkel cells.

Convergence of Sensory Fibers to Merkel Cells in Hair Follicles

To examine the relationship between Merkel cells and sensory fibers that grew in a manner converging to the hair follicle, live Merkel cells in a co-culture preparation were identified by FM1-43 dye. In this co-culture experiment, Merkel cells in a hair follicle were stained with FM1-43 in advance of the culture. When a co-culture preparation of a hair follicle and a DRG for 3 days in collagen gel with NGF and NT3 was examined under a phase-contrast view, many nerve fibers were seen to converge (open arrows) onto the superior enlargement of the hair follicle. When the same preparation was then examined under a fluorescence microscope, FM-143-containing Merkel cells were seen forming groups in several spots (filled arrows) on one side of the follicular preparation. In Figure 6a phase-contrast and fluorescence observations are superimposed. Convergence of sensory fibers can be

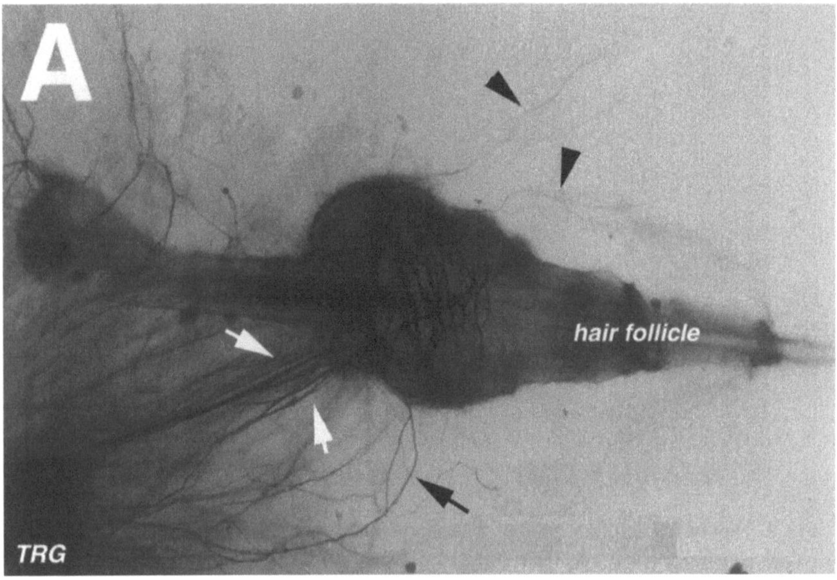

Fig. 4A, B. Three-dimensional view of distribution of synaptophysin 1 around a hair follicle innervated by DRG fibers during co-culture in collagen gel. A co-culture preparation of a hair follicle and a DRG in collagen gel was immunostained against synaptophysin 1 and was colored by Cy3 conjugated with the secondary antibody. **A** Lateral view of an immunostained hair follicle. Synaptophysin 1 immunostaining is enriched around the superior enlargement of the preparation (*arrow*). **B** Cross section of the hair follicle in **A** by means of a computer. Synaptophysin 1 immunostaining can be seen in the surface of the hair follicle preparation

seen to the places at which the FM-dye-stained Merkel cells were enriched. Accordingly, it was highly likely that Merkel cells in the hair follicles attracted sensory nerve fibers.

Discussion

The present co-culture study of sinus hair follicles and sensory ganglia in collagen gel demonstrates that the hair follicles represent the capability to attract sensory nerve fibers during incubation with serum-free medium containing NGF and NT3. The presence of a large number of Merkel cells in the region where the sensory nerve fibers made convergence growth was demonstrated by immunostaining against cytokeratin 20. It was also demonstrated that a synaptic-specific protein, synaptophysin 1, is enriched in nerve fibers that grow circling the superior enlargement. It was thus likely that synaptic structures were formed by newly grown sensory nerve fibers around the Merkel cell-rich areas of the hair follicles in the co-culture preparations.

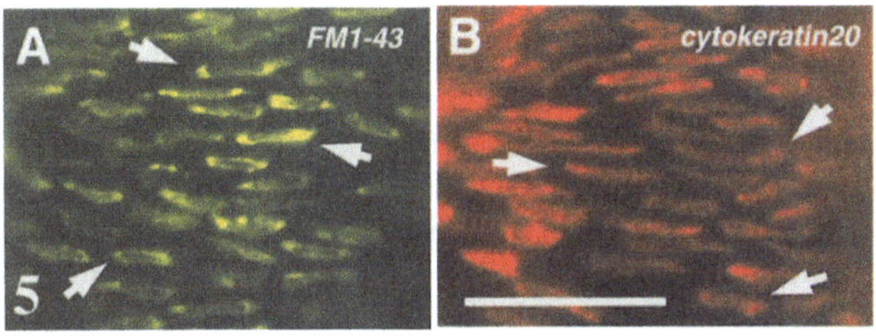

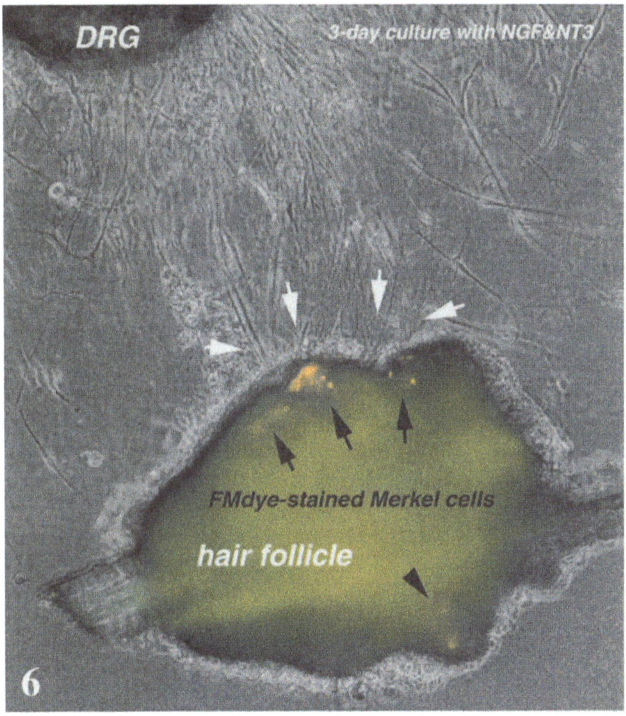

Fig. 5A, B. Identification of Merkel cells in hair follicles by means of vital staining with FM1-43 and immunostaining against cytokeratin 20. **A** Merkel cells of a hair follicle of a 9-day-old rat stained vitally with FM1-43. FM-dye-stained Merkel cells were seen as spindle-like cells (*arrows*). **B** Merkel cells in a hair follicle of a 5-day-old rat immunostained against cytokeratin 20 and colored by Cy3. They were seen as spindle-like cells indicated by *arrows*

Fig. 6. Convergence of DRG fibers to FM1-43-stained Merkel cells in a hair follicle. A hair follicle and a DRG were co-cultured in collagen gel with serum-free medium supplemented with NGF and NT3 for 3 days. Merkel cells in the hair follicle were stained vitally with FM1-43 prior to co-culture. Under a phase-contrast microscope, nerve fibers from the DRG were seen to converge (*white arrows*) on the hair follicle. When the hair follicle was examined under a fluorescence microscope, FM1-43-stained Merkel cells were clustering (*black arrows*) at several positions to which sensory nerve fibers converged. Some FM1-43-stained Merkel cells were also seen located on the opposite side of the hair follicle (*black arrowhead*), to which no DRG fibers had grown

Involvement of Merkel cells in forming convergence growth of sensory nerve fibers to the superior enlargement was strongly suggested in our co-culture. The regions of the hair follicles to which the sensory nerve fibers converged were demonstrated to be enriched with Merkel cells by immunostaining against cytokeratin 20. The convergence of sensory nerve fibers was still observed even after removal of connective tissues surrounding the hair follicle. Moreover, when Merkel cells were labeled by an FM dye in advance of co-culture, cells that contained the fluorescence dye were seen to aggregate in the areas to which sensory nerve fibers made the convergence type of growth. These observations strongly suggest that Merkel cells have a function of attracting sensory nerve fibers.

References

Andres KH (1966) Über die Feinstruktur der Rezeptoren an Sinushaaren. Z Zellforsch 75:339–365

Fukuda J (1996) A pure, monolayer culture of Merkel cells from sinus hair follicles of the rat. Neurosci Lett 216:73–76

Lumsden AG, Davies AM (1986) Chemotropic effect of specific target epithelium in the developing mammalian nervous system. Nature 323:538–539

Ryan TA (2001) Presynaptic imaging techniques. Curr Opin Neurobiol 11:544–549

Sudhof TC (2000) The synaptic vesicle cycle revisited. Neuron 28:317–320

Tachibana T, Fujiwara N, Nawa T (2000) Postnatal differentiation of Merkel cells in the rat palatine mucosa, with special reference to the timing of peripheral nerve development and the potency of cell mitosis. Anat Embryol 202:359–367

Vos P, Stark F, Pittman RN (1991) Merkel cells in vitro: production of nerve growth factor and selective interactions with sensory neurons. Dev Biol 144:281–300

Merkel Cell Development is Independent of L1 Adhesion Molecule

D. Magnani , S. Kief, J.M. Brandner, U. Bartsch, M. Schachner, I. Moll

Department of Dermatology and Venerology, University Hospital Hamburg-Eppendorf, Hamburg, Germany

Summary

Merkel cells express epithelial and neuronal markers. In the skin of snout and footpad of L1 knock-out and wild-type mice, similar densities of Merkel cells were observed by immunohistochemistry. This result indicates that Merkel cells are independent of L1 presence.

Introduction

L1 adhesion molecule (L1) is a 200–220 kDa type I membrane glycoprotein of the immunoglobulin superfamily expressed in a variety of neural and epithelial cells. It supports homophilic as well as integrin-mediated cell binding. Besides cell adhesion, it is involved in cell migration, and cell growth especially of neuronal cells. It has been shown to deliver signals for the growth and survival of neuronal cells.

The neuroendocrine cells of mammalian skin, the Merkel cells (MC) are supposed to be epithelially derived, but they show both epithelial and neuronal markers. They express cytokeratins (nos. 8, 18, 19, 20), desmosomal proteins and chromogranin A, neuron-specific enolase and various neuropeptides. Because of the importance of L1 for both epithelial and neuronal cells, we were interested in the existence of MC in L1 knock-out mice in comparison to wild-type mice.

Material and Methods

The skin of the snout and footpad of L1 knock-out and wild-type mice were frozen immediately after preparation and 6-μm-thick cryostat sections were made. MC were localized by using antibodies against cytokeratin 18 and 20 (Fig. 1), as

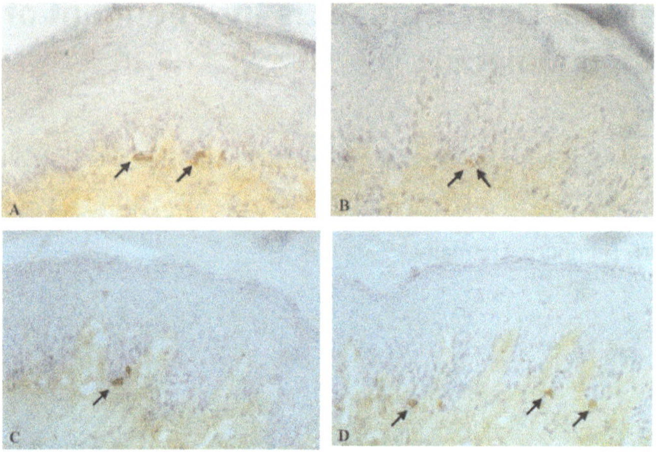

Fig. 1. Indirect immunostaining on an acetone-fixed cryosection of a murine foot pad using monoclonal antibodies against CK18 (**A**, **B**) and CK20 (**C**, **D**) comparing wild-type mice (**A**, **C**) and L1 knock-out mice (**B**, **D**). *Arrows* MC. Note the presence of MC in both wild-type and L1 knock-out skin

described by Moll et al. (1996), counted and estimated per mm^2 of frozen section area. Primary antibodies were kindly provided by W.W. Franke and coworkers, German Cancer Research Center, Heidelberg. Tissues of L1 knock-out mice were kindly provided by M. Schachner and coworkers, ZMNH, Hamburg (cf. Dahme et al. 1997).

Results and Discussion

We could identify MC in knock-out and wild-type mice with identical density in the foot pad (Fig. 2, left columns). The higher number of MC in the snout of wild-type mice compared to knock-out mice was not significant (Fig. 2, right columns).

The very similar numbers of MC in the skin of L1 knock-out and of wild-type mice strongly argue for a regular fetal development of MC in the absence of L1.

L1 provided by nerves may not be a prerequisite for their development. This is in accordance with former data which have shown the presence of MC in murine epidermis clearly before cutaneous nerves are present (Pasche et al. 1990).

The same is true for human epidermis; MC are detectable around fetal week 8 without any cutaneous nerves in their neighbourhood (Moll et al. 1996). However, it still remains to be clarified whether L1 is important for homeostasis and function of MC and their synapses in later life. This shall be elucidated in future experiments.

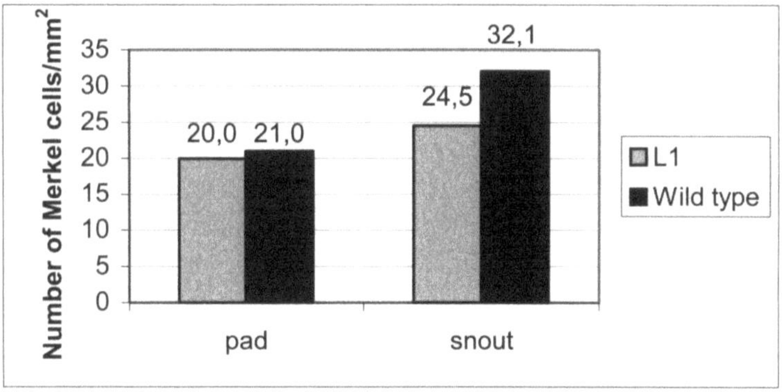

Fig. 2. Number of Merkel cells/mm^2 in snout and foot pad of L1 knock-out mice in comparison to wild-type mice. MC were localized by antibodies against CK20, counted and estimated per mm^2. We identified MC in both wild-type and knock-out mice. The numbers of MC in foot pads of wild-type and knock-out mice were similar. The higher number of MC/mm^2 in the snout of wild-type in comparison to L1 knock-out mice was not significant

References

Dahme M, Bartsch U, Martini R, Anliker B, Schachner M, Mantei N (1997) Disruption of the mouse L1 gene leads to malformations of the nervous system. Nat Genet 17:346–349

Moll I, Paus R, Moll R (1996) Merkel cells in mouse skin: intermediate filament pattern, localization, and hair cycle-dependent density. J Invest Dermatol 106:281–286

Pasche F, Mérot Y, Carraux P, Saurat J-H (1990) Relationship between Merkel cells and nerve endings during embryogenesis in the mouse epidermis. J Invest Dermatol 95:247–251

Merkel Cell

Mechanoreceptors

The Functional Significance of the Vibrissal System of Marine Mammals

Guido Dehnhardt[1], Björn Mauck[1] and Heikki Hyvärinen[2]

[1]Department of General Zoology and Neurobiology, University of Bochum, 44780 Bochum, Germany

[2]Department of Biology, University of Joensuu, 80101 Joensuu, Finland

Summary

While the structure of the vibrissal follicle-sinus complex (FSC) is best described in terrestrial mammals, most of our understanding of the functional significance of vibrissae stems from studies on marine mammals. This review focuses on functional aspects of the conspicuously well innervated vibrissae of pinnipeds that enable the animals to forage under conditions where sight is reduced. As in rats and sirenians, the vibrissae of pinnipeds are designed for the reception of tactile information, which the animals receive through physical contact of the hairs to an object. Using this so-called haptic sense, the animals are capable of identifying the size, shape and surface structure of an object. The performance of pinnipeds on such tasks is comparable to the prehensile hands of some primate species and may help them when foraging at the sea bottom. Further studies demonstrated that the vibrissae of seals are also highly sensitive to minute water movements and thus form a hydrodynamic receptor system like the lateral line of fish. Contrary to the accepted view that hydrodynamic object detection works only over short distances, these studies have shown that seals can locate distant objects by hydrodynamic trail-following. These results establish a new system for spatial orientation in the aquatic environment that can explain successful feeding of pinnipeds in dark and murky waters.

Appearance of Vibrissae in Marine Mammals

The occurrence and distribution of vibrissae or sinus hairs in marine mammals is diverse (Ling 1977). Baleen whales, for instance, possess about 100 very thin vibrissae (ca. 0.3 mm in diameter) around the blowholes and along the lower and/

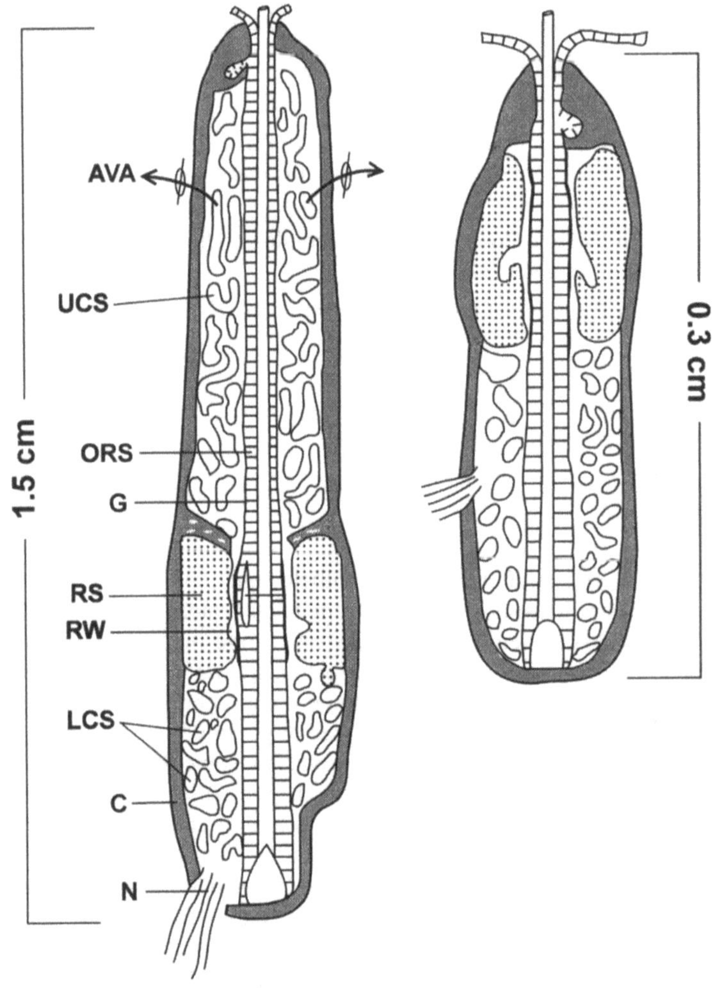

Fig. 1. Schematic diagram of a longitudinal section of a mystacial vibrissae of the ringed seal (*left*) and the rat (*right*). *A* Arterio-venous anastomoses, *UCS* upper cavernous sinus, *ORS* outer root sheath, *G* glassy membrane at the surface of the basal cell layer of the outer root sheath, *RS* ring sinus, *RW* ringwulst, *LCS* lower cavernous sinus, *N* nerve bundle penetrating the capsule (*C*)

or upper jaw. Most odontocetes lose these tactile hairs postnatally (Ling 1977), whereas species belonging to the order Sirenia possess vibrissae all over the body (Reep et al. 1998, 2001). In pinnipeds, the facial vibrissae can be divided into the mobile mystacial vibrissae on the muzzle, the immobile supraorbital vibrissae (above each eye), and rhinal vibrissae (only in phocid seals) vertically situated on the back of the muzzle.

With respect to the structure of the hair shaft, the vibrissae of phocid seals show an interesting differentiation. While vibrissal hair shafts of the bearded seal (*Erignathus barbatus*) and the monk seals (*Monachus* spp.) are oval in diameter and smooth in outline, those of all other phocid species are extremely flattened and have waved surfaces (Watkins and Wartzok 1985; Hyvärinen 1989; Dehnhardt and Kaminski 1995). These differences in hair structure are suggested to be mechanical adaptations to the signals the hairs receive and transmit to the receptors within the hair follicle (Dehnhardt and Kaminski 1995).

Structure and Innervation of Vibrissal Follicles in Pinnipeds

Although the basic morphology of the vibrissal follicle-sinus complex ("FSC", Rice et al. 1986) of pinnipeds compares well to that of FSCs of most terrestrial mammals, differences in FSC size, the structure of the sinus system as well as the degree and pattern of innervation have been described, which may reflect differences in sensitivity and function.

The first apparent and striking feature of the FSCs of pinnipeds is their remarkable size (Fig. 1). A comparison of pinniped FSCs (which can measure up to 2 cm) with those of terrestrial mammals (e.g., in the leopard *Panthera pardus*, a species of comparable body size, FSCs are 6.5 mm long, Dehnhardt et al. 1999) suggests that FSC size is not merely determined by body size, but has to be considered as a morphological parameter of its own adaptive significance.

In contrast to the bipartite blood sinus of most terrestrial mammals (Andres 1966) composed of a ring sinus close to the apical end of the follicle and a cavernous sinus located below it, the FSCs of pinnipeds (*Zalophus californianus*, Stephens et al. 1973; *Phoca hispida*, Hyvärinen and Katajisto 1984; Hyvärinen 1989) possess an upper cavernous sinus that takes about 60% of the total length of the follicle (Fig. 1). The ring sinus area, where most mechanosensitive receptors are located, is inserted much deeper in the capsule than in terrestrial mammals.

At the level where the deep vibrissal nerve is passing through the capsule, the number of nerve fibres of the ringed seal (*Phoca hispida*, Hyvärinen and Katajisto 1984; Hyvärinen 1989) exceeds that calculated for terrestrial species like the rat (Rice et al. 1986) by a factor of 10 (1600 vs. 160). This difference in innervation density is the same for the different mechanoreceptors located in the area of the ring sinus (Hyvärinen 1989; Halata 1993). Although the number of Merkel cell-neurite complexes (MCs, 10–20,000 per FSC) makes it clear that this is by far the dominating sensory element, there are also encapsulated end-organs (lanceolate endings, 1000–4000 per FSC; lamellated endings, 100–400 per FSC) and numerous small free nerve endings in the ring sinus and the lower cavernous sinus area.

In the entire upper cavernous sinus, no sensory elements can be found. From the level of the ringwulst to the beginning of the upper cavernous sinus, MCs are situated below the glassy membrane in the basal cell layer of the outer root sheath (Fig. 2). All of these MCs are contacted by nerve terminals.

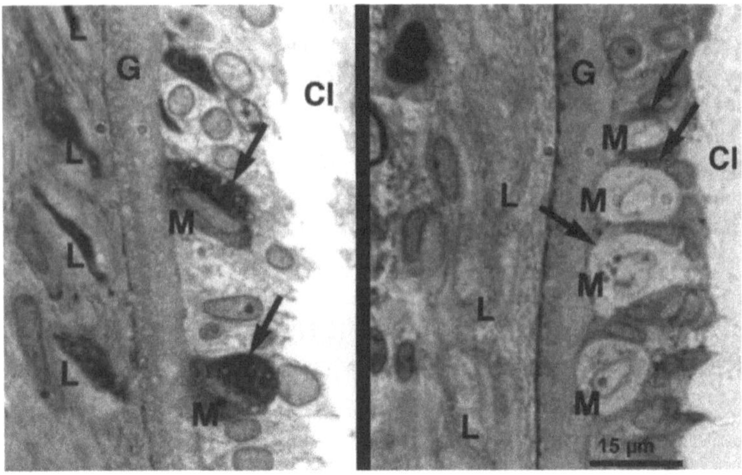

Fig. 2. Light micrographs of the ring sinus area. Merkel endings are situated below the glassy membrane (*G*) in the basal cell layer of the outer root sheath, while the lanceolate nerve endings (*L*) are on the surface of the glassy membrane. The liquid filled cleft (*C*) lines the Merkel endings. *Left* Type 1 Merkel cells (*M*) are quite narrow. *Right* Type 2 Merkel cells are large and rounded. The nerve endings are indicated by *arrows*

The Functional Significance of Mystacial Vibrissae

The functional significance of vibrissae remained obscure for a long time. Results from single unit recordings at the infraorbital branch of the trigeminal nerve of harbor seals and grey seals suggested that these tactile hairs are primarily designed for the reception of tactile information which the animals receive through physical contact of the vibrissae to objects in the environment (Dykes 1975). Such active touch achievements are a function of the so-called haptic sense, integrating cutaneous mechanosensation as well as kinaesthetic information (Gibson 1962; Loomis and Lederman 1986).

The different types of facial vibrissae of manatees and dugongs form a haptic system unique among marine mammals. Manatees use their vibrissal-muscular complex, consisting of the rigid perioral vibrissae and the muscular lips, to grasp and bring food into the oral cavity when feeding on submerged vegetation (Marshall et al. 1998). Bachteler and Dehnhardt (1999) demonstrated that manatees use the thin bristle-like vibrissae of the oral disc for the discrimination of textured surfaces. This is consistent with the results of Marshall et al. (1998), suggesting that the bristle-like hairs of the oral disc are involved in the tactile exploration of objects.

In accordance with the neurophysiological results obtained by Dykes (1975), psychophysical experiments have shown that, similar to manatees, pinnipeds are capable of identifying the size, shape and surface structure of an object by active

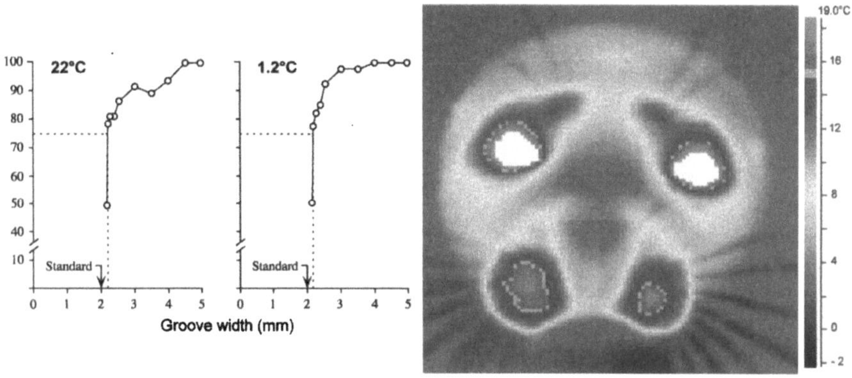

Fig. 3. Psychometric functions of performance of two harbor seals discriminating grooved surfaces by active touch. The seals were required to choose the standard stimulus (*vertical arrow*) in each stimulus combination. The *vertical dashed line* indicates the size of the interpolated comparison stimulus at threshold (75% correct choices). Each data point represents the result of at least 40 decisions. The infrared thermogram shows the typical distribution of temperatures measured on the surface of a seal's face immediately after the animal had left water of approximately 1 °C

touch with their mystacial vibrissae (Kastelein and van Gaalen 1988; Dehnhardt 1990 1994; Dehnhardt and Kaminski 1995; Dehnhardt and Dücker 1996; Dehnhardt et al. 1997, 1998a). In the California sea lion, tactile shape recognition is as fast and reliable as by vision. Vibrissal size discrimination capabilities in harbor seals are comparable to those of the prehensile hands of some primate species (Carlson et al. 1989) and comes close to the visual resolving power of pinnipeds.

In their aquatic environment marine mammals allow their outermost tissue layers to cool down close to ambient temperature (Kvadsheim et al. 1997). While in humans a decrease in skin temperature leads to severe deterioration of tactile sensitivity (Green et al. 1979; Gescheider et al. 1997), the ability of harbor seals to discriminate textured surfaces remains unaltered under different thermal conditions (Dehnhardt et al. 1998a; Fig. 3). Even at water temperatures of about 1 °C, the seals achieved Weber fractions of 0.09. Infrared thermography revealed that, contrary to the dictate of thermal economy, the mystacial and supraorbital vibrissal pads of harbor seals are areas of excessive heat loss (Fig. 3). Thermally clearly defined against the rest of the head, the high temperatures measured at the surfaces of vibrissal pads vividly demonstrate that in these sensory areas no vasoconstriction occurs during cold acclimation, indicating a separate vibrissal blood circulation. Selective heating of vibrissal pads is suggested to be a function of the upper cavernous blood sinus (see Fig. 1), which is free of receptors and thus may primarily serve as a thermic insulator for the receptor area below it (Mauck et al. 2000).

Observations in the wild suggest that sea lions (*Otaria byronia*) and walruses use the haptic function of their vibrissae when foraging at the sea bottom (Lindt 1956; Fay 1982). The question remains whether vibrissae could provide sensory

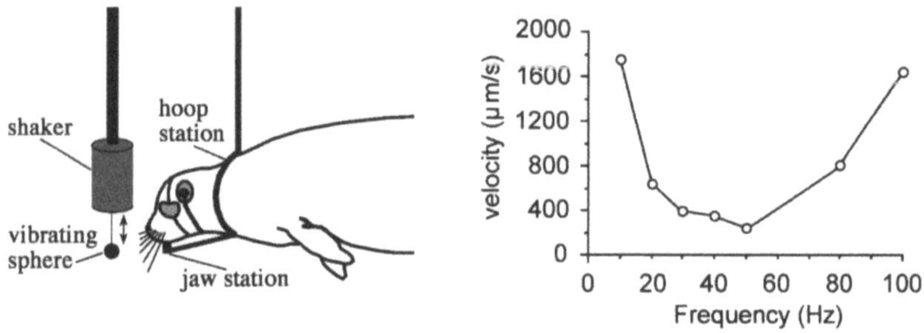

Fig. 4. Detection of water movements by the harbor seal. At the beginning of a trial the hoop station was above the water surface, where the seal was stationed and supplied with eye caps and head phones, or, during whisker-exclusion tests, with a muzzle of wire mesh. Then the hoop was submerged with the seal to the final test position. The seal was trained to correct the position of its lower jaw whenever it lost contact with the knob of the jaw station. The behavioral displacement thresholds (50% correct decisions) in terms of the velocity of sinusoidal water movements are typical for a hydrodynamic receptor system

information for the detection of pelagic fish. In the aquatic environment, one utilizable source of sensory information consists of water disturbances, inevitably caused by any moving organisms. Hydrodynamic sensory systems, like the lateral line of fish, have evolved many times in aquatic animals (Bleckmann 1994). That the vibrissae of harbor seals also function as a hydrodynamic receptor system has been demonstrated with a technique commonly used to study the fish lateral line (Dehnhardt et al. 1998b). Water movements (10–100 Hz) were generated with a constant volume oscillating sphere positioned 5–50 cm in front of the vibrissae of a harbor seal (Fig. 4). The shape of the tuning curve obtained for a harbor seal (Fig. 4) is similar to those determined for other aquatic animals equipped with a hydrodynamic sense (Bleckmann 1994), and characterizes the vibrissae as a hydrodynamic receptor system with a spectral sensitivity well tuned to the frequency range of fish-generated water movements.

The accepted view is that hydrodynamic object detection works only over short distances as, for instance, during the final stage of prey pursuit of seals. Particle velocities attenuate rapidly with distance from the flow field-generating source. However, the wake behind a swimming fish shows a vortex structure (Fig. 5) with particle velocities above the threshold of most hydrodynamic receptors several minutes after the fish has passed by. Thus, a swimming fish can leave a hydrodynamic trail of considerable length that a piscivorous predator might use for long-range prey detection Hanke et al. (2000). To find out whether seals can locate distant objects by hydrodynamic trail-following, Dehnhardt et al. (2001) used a miniature submarine for the generation of hydrodynamic trails and visualized and measured them by particle image velocimetry (PIV). Measurements showed that the submarine's trail was a narrow street of turbulent water movements containing water velocities in the same order of magnitude as velocities calculated for the wake of a fish of 30-cm body length. Linear hydrodynamic trails as well as trails

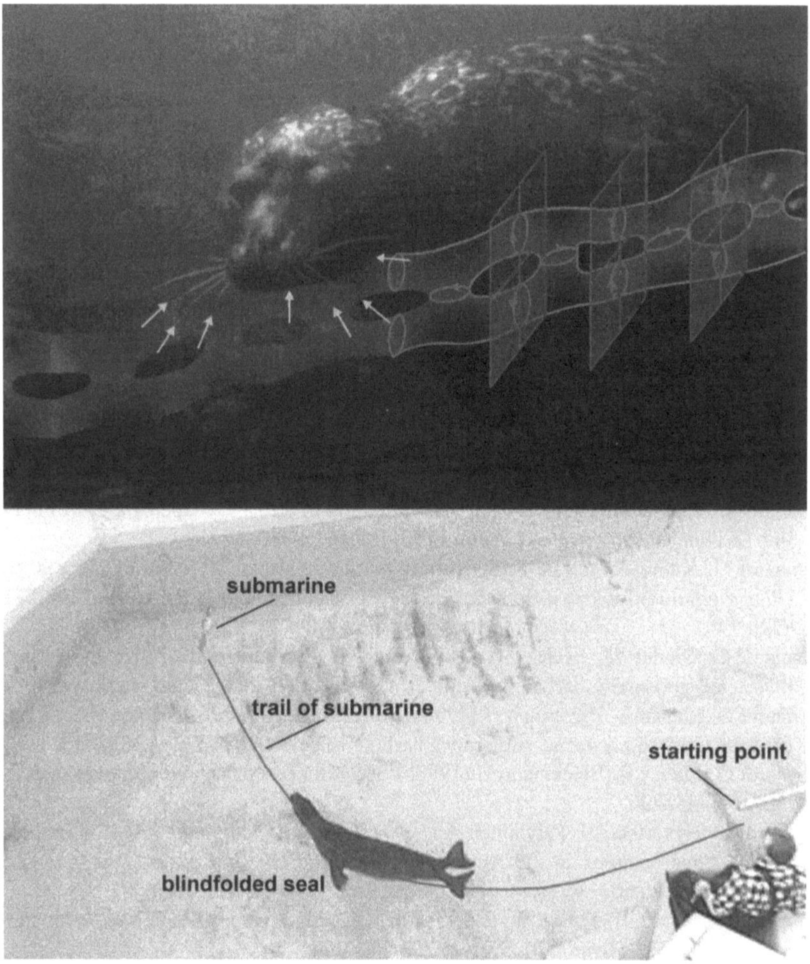

Fig. 5. The arrangement of mystacial vibrissae allows this ringed seal to perform multiple point measurements of the water movements in the wake of a swimming fish. This way it can obtain information about the three-dimensional vortex structure of water movements. The single frame taken from video recordings shows a trial with a hydrodynamic trail containing a right-hand curve

containing unpredictable changes were of course generated. When the blindfolded seal was signaled to search for the submarine, it immediately submerged and protracted the vibrissae to the most forward position. The animal turned onto the submarine's course as soon as it intersected the hydrodynamic trail and exactly followed it to the final position of the submarine. These results establish a new system for spatial orientation in the aquatic environment that might help explain successful feeding of pinnipeds in dark and murky waters.

References

Andres KH (1966) Über die Feinstruktur der Rezeptoren an Sinushaaren. Z Zellforsch 75:339–365

Bachteler D, Dehnhardt G (1999) Active touch performance in the Antillean manatee: evidence for a functional differentiation of facial tactile hairs. Zoology 102:61–69

Bleckmann H (1994) Reception of hydrodynamic stimuli in aquatic and semi-aquatic animals. In: Rathmayer W (ed) Progress in Zoology 41. Fischer, Stuttgart

Carlson S, Tanila H, Linnankoski I, Pertovaara A, Kehr A (1989) Comparison of tactile discrimination ability of visually deprived and normal monkeys. Acta Physiol Scand 135:405–410

Dehnhardt G (1990) Preliminary results from psychophysical studies on the tactile sensitivity in marine mammals. In: Thomas JA, Kastelein RA (eds) Sensory abilities of cetaceans. Plenum Press, New York, pp 435–446

Dehnhardt G (1994) Tactile size discrimination by a California sea lion (*Zalophus californianus*) using its mystacial vibrissae. J Comp Physiol 175:791–800

Dehnhardt G, Dücker G (1996) Tactual discrimination of size and shape by a California sea lion (*Zalophus californianus*). Anim Learn Behav 24:366–374

Dehnhardt G, Kaminski A (1995) Sensitivity of the mystacial vibrissae of harbour seals (*Phoca vitulina*) for size differences of actively touched objects. J Exp Biol 198:2317–2323

Dehnhardt G, Sinder M, Sachser N (1997) Tactual discrimination of size by means of mystacial vibrissae in harbour seals: in air versus underwater. Z Säugetierkd 62:40–43

Dehnhardt G, Mauck B, Hyvärinen H (1998a) Ambient temperature does not affect the tactile sensitivity of mystacial vibrissae in harbour seals. J Exp Biol 201:3023–3029

Dehnhardt G, Mauck B, Bleckmann H (1998b) Seal whiskers detect water movements. Nature 394:235–236

Dehnhardt G, Hyvärinen H, Palviainen A (1999) Structure and innervation of the vibrissal follicle-sinus complex in the Australian waterrat (*Hydromys chrysogaster*). J Comp Neurol 411:550–562

Dehnhardt G, Mauck B, Hanke W, Bleckmann H (2001) Hydrodynamic trail-following in harbor seals (*Phoca vitulina*). Science 293:102–104

Dykes RW (1975) Afferent fibres from mystacial vibrissae of cats and seals. J Neurophysiol 38:650–662

Fay FH (1982) Ecology and biology of the Pacific walrus, *Odobenus rosmarus divergens* Illiger. Fish and Wildlife Service Publication No. 74. Fish and Wildlife Service, Washington, DC

Gescheider GA, Thorpe JM, Goodarz J, Bolanowski SJ Jr (1997) The effects of skin temperature on the detection and discrimination of tactile stimulation. Somatosens Mot Res 14:181–188

Gibson JJ (1962) Observations on active touch. Psychol Rev 69:477–491

Green BG, Lederman SJ, Stevens JC (1979) The effect of skin temperature on the perception of roughness. Sensory Processes 3:327–333

Halata Z (1993) Sensory innervation of the hairy skin (light- and electron microscopic study). J Invest Dermatol 101:75–81

Hanke W, Brücker C, Bleckmann H (2000) The ageing of the low-frequency water disturbances caused by swimming goldfish and its possible relevance to prey detection. J Exp Biol 203:1193–1200
Hyvärinen H (1989) Diving in darkness: whiskers as sense organs of the ringed seal (*Phoca hispida*). J Zool 218:663–678
Hyvärinen H, Katajisto H (1984) Functional structure of the vibrissae of the ringed seal (*Phoca hispida*, Schr.). Acta Zool Fennica 171:27–30
Kastelein RA, van Gaalen MA (1988) The sensitivity of the vibrissae of a Pacific walrus (*Odobenus rosmarus divergens*). Aquat Mammals 14:123–133
Kvadsheim PH, Gotaas ARL, Folkow LP, Blix AS (1997) An experimental validation of heat loss models for marine mammals. J Theor Biol 184:15–23
Lindt CC (1956) Underwater behaviour of the southern sea lion, *Otaria jubata*. J Mammal 37:287–288
Ling JK (1977) Vibrissae of marine mammals. In: Harrison RJ (ed) Functional anatomy of marine mammals, vol 3. Academic Press, London, pp 387–415
Loomis JM, Lederman SJ (1986) Tactual perception. In: Boff KR, Kaufman L, Thomas JR (eds) Handbook of perception and human performance, vol 2. Wiley, New York, pp 1–41
Marshall CD, Huth GD, Edmonds VM, Halin DL, Reep RL (1998) Prehensile use of perioral bristles during feeding and associated behaviours of the Florida manatee (*Trichechus manatus latirostris*). Mar Mammal Sci 14:274–289
Mauck B, Eysel U, Dehnhardt G (2000) Selective heating of vibrissal follicles in seals (*Phoca vitulina*) and dolphins (*Sotalia fluviatilis guianensis*). J Exp Biol 203:2125–2131
Reep RL, Marshall CD, Stoll ML, Whitaker DM (1998) Distribution and innervation of facial bristles and hairs in the Florida manatee (*Trichechus manatus latirostris*). Mar Mammal Sci 14:257–273
Reep RL, Stoll ML, Marshall CD, Homer BL, Samuelson DA (2001) Microanatomy of facial vibrissae in the Florida Manatee: the basis for specialized function and oripolation. Brain Behav Evol 58:1–14
Rice FL, Mance A, Munger BL (1986) A comparative light microscopic analysis of the innervation of the mystacial pad. I. Vibrissal follicles. J Comp Neurol 252:154–174
Stephens RJ, Beebe IJ, Poulter TC (1973) Innervation of the vibrissae of the California sea lion, *Zalophus californianus*. Anat Rec 176:421–442
Watkins WA, Wartzok D (1985) Sensory biophysics of marine mammals. Mar Mammal Sci 1:219–260

Confocal Microscopic Analysis of Merkel Innervation in the Cat Mystacial Vibrissa Follicles

Satomi Ebara[1], Kenzo Kumamoto[1] and Frank L. Rice[2]

[1]Department of Anatomy, Meiji University of Oriental Medicine, Kyoto 629-0392, Japan
[2]Center for Neuropharmacology and Neuroscience, Albany Medical College, Albany, New York 12208, USA

Summary

Three-dimensional morphology of the Merkel innervation in very thick sections of the cat mystacial vibrissa follicles as well as in whole mounted tissues were demonstrated by using a confocal laser microscope with a powerful analyzing system for reconstruction. A sheet of densely arranged Merkel nerve endings usually having scattered gaps, was observed at the level of the ring sinus around a large sized follicle. The number of all Merkel nerve terminal disks was roughly estimated at about 3200 in the follicle. The gaps may suggest plasticity among the Merkel innervation.

Results and Discussion

Vibrissa follicle-sinus complexes (FSCs) are exceptionally well-innervated, important tactile organs on the mystacial pads of most mammals (Rice et al. 1993). At least eight distinct endings including two kinds of Merkel innervation, at the mouth of the follicle and at the level of the ring sinus were clearly discriminated in the rat and cat FSCs by using confocal laser scanning microscopy (CLSM; Ebara et al. 2002). Recently, we developed a whole mount technique in order to observe the full innervation around the FSCs. In this study, we focused on the ring sinus Merkel innervation in the cat large FSCs and demonstrated the three-dimensional (3D) structures by using mainly the anti-protein gene product 9.5 antiserum as neuronal markers. Abundant CLSM stack images taken from the whole mounted FSCs by rotating around the shaft as well as from the 100–200 μm-thick sections were made into 3D reconstructions by using a worldwide-first highly sophisticated application, VG StudioMax 1.1 software (Volume Graphics, Germany). We could easily analyze and understand the detailed

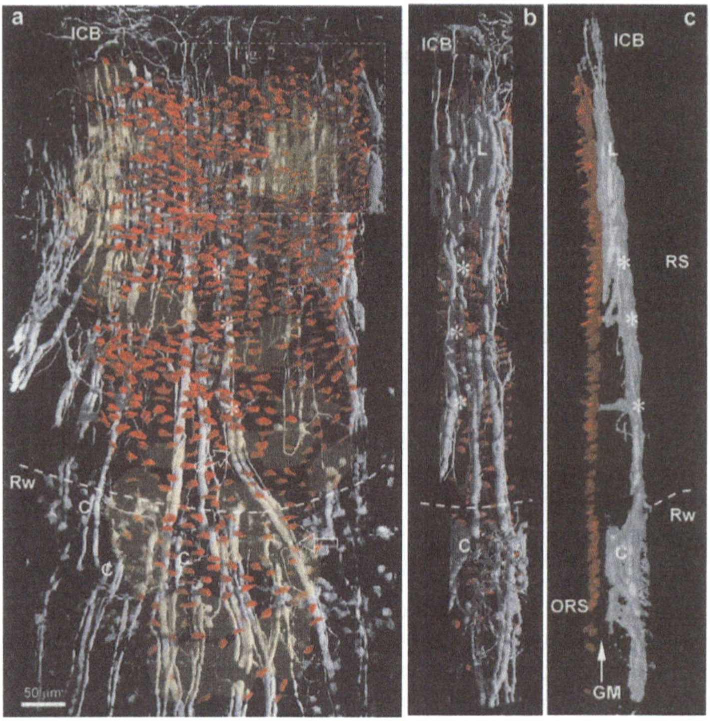

Fig. 1a–c. PGP immunopositive innervation in a 150-μm-thick section of cat FSC (#2 of row D). Merkel nerve endings situated in the outer root sheath were pseudo-colored in red. **a** View from the side of the vibrissal shaft. **b** Back view of the middle part of **a**. **c** Left side view of **b**. A large caliber nerve fiber branches twice (*open arrows*), each branch penetrates the glassy membrane (*GM*) and terminates as the Merkel terminal disks (*asterisks*). Distorted and/or faintly labeled endings result in gaps in the total distribution of the endings (*yellow areas*; cf. Fig. 2). Including these irregular endings, the number of all endings was about 592 in **a**. The *dashed line* indicates the position of the ring wulst (*Rw*). *L* Lanceolate endings, *C* club-like ring-wulst endings

morphology in spite of the highly complicated structure of innervation that is distributed in a very thin layer of the follicle (external root sheath) which is overlaid by a layer of dense lanceolate innervation in the mesenchymal sheath.

In three dimensions, we could discriminate and segment all Merkel nerve endings that are situated in the external root sheath in both images taken from a thick sectioned FSC and a whole mounted FSC (Figs. 1, 3). The distal ends of the Merkel terminal nerve fibers ascending in the mesenchymal sheath penetrated the glassy membrane (GM) to form mushroom cap-like terminal disks on Merkel cells. Merkel cells were also labeled with anti-PGP, but usually appeared very faintly. Morphologically remarkable differences of Merkel innervation at the level of the ring sinus between different species were suggested to have different functional properties (Figs. 1, 2; cf. Ebara et al. 2002).

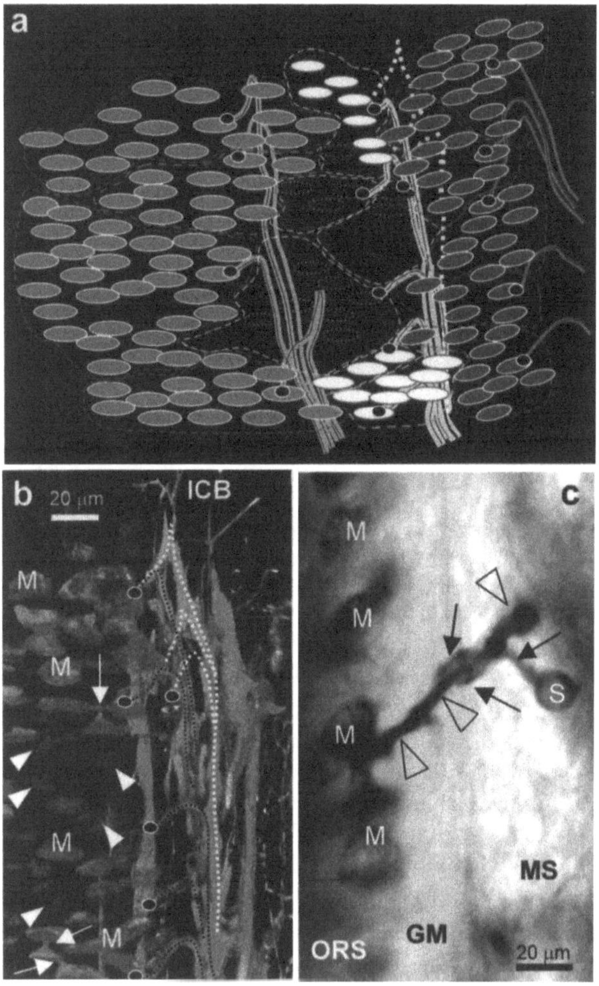

Fig. 2. **a**, **b** Reconstruction of Merkel innervation of *dashed square* in Fig. 1a. **b** A projection view of several identical slice images of the middle portion of **a**. Each Merkel terminal nerve fiber (*solid lines* in **a** and *black dotted lines* in **b**) penetrates the *GM* (*black small circles*) and provides a grid-like territory (*dashed contours* in **a**) containing 15–30 Merkel terminal disks (*M* in **b**, *ovals* in **a**) connected in series by thin processes (*white arrows*). In contrast to haphazard terminations in the rat, the Merkel nerve fibers in the cat FSC were usually en passant branches whose terminal grids were immediately adjacent to each other along the long axis of the follicle. Distorted or faintly labeled Merkel cell-nerve endings (*white arrow heads* in **b**, *dotted ovals* in **a**) are observed all together in small areas. Interestingly, the second kind of fibers were observed approaching the irregular areas (*white dotted lines*). **c** Anti-PGP with DAB immunocytochemistry in a 200-µm-thick section of the cat FSC. The Merkel afferent (*open arrowheads*) is surrounded by cell processes originating from the cell body of the podocyte-like terminal Schwann cell (*S*)

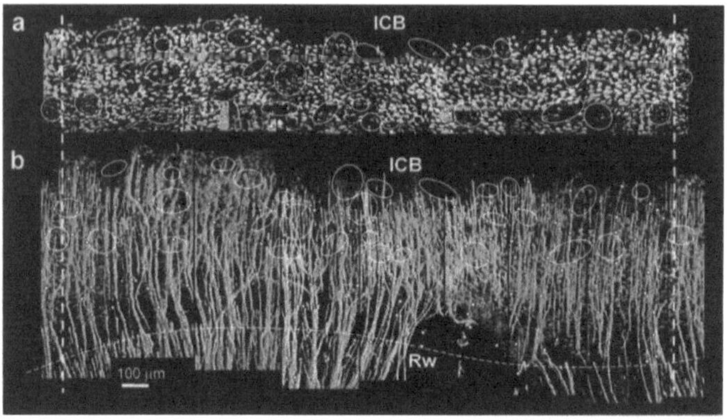

Fig. 3a, b. A PGP-immuno-labeled full innervation at the level of the ring sinus around the cat large FSC (#1 of row D). **b** depicts eight confocal images taken from a whole mounted FSC by using a 20× objective lens. **a** is the result of segmentation and montage of about 20 images (60×) of only Merkel nerve endings of the upper third of **b**. About 1630 endings were in **a**. *Longitudinal dashed lines* indicate the same position of the FSC. The ring wulst (*Rw*) is situated below the *curved dashed line*. *Ovals* Irregular areas of Merkel nerve endings at the upper level of the RS, *small white circles* in **b**; terminal points of the Merkel nerve fibers adjacent to the GM

The density of the Merkel nerve endings in a thick section of large sized FSC (#2D) is 0.30 endings per 100 μm^2 (Fig. 1). At a rough estimate, the density was about three times higher at the upper third of the ring sinus than the lower third, namely the level of the ring wulst. Although some endings were present, nerve fibers scarcely terminated to the lower level (Figs. 1, 3). By using a whole mount technique, we could see almost all of the Merkel innervation surrounding the whole follicle of another large FSC (#1D;(ig. 3). By counting all terminal disks at the level of the upper third of the ring sinus (density; 0.25 per 100 μm^2 uniformly in the area), about 3200 endings were roughly estimated around the follicle. This is consistent with a previous study that estimated the number of Merkel cells at about 3000, even in a small FSC (Gottschaldt and Vahle-Hinz 1981). However, they only observed that 50–70% of the cells were innervated.

Interestingly, we routinely found scattered sites around the follicle where one or two kinds of fiber approached or penetrated the GM, but only formed a few distorted and/or faintly labeled endings resulting in gaps in the total distribution of Merkel disks around the follicle (Figs. 1, 2a,b, 3). The gaps were scattered on the follicle. Several gaps had a tendency to be arranged along the long axis of the follicle. These observations may suggest a plasticity among the Merkel innervation. Surprisingly, at least in the cat, some unique terminal Schwann cells had thin processes resembling podocytes that sheath the terminal tips of Merkel afferent fibers (Fig. 3). Almost all Merkel afferent terminal fibers in the cat were recognized to be surrounded by a few terminal Schwann cells that clearly were S100-immuno-positive.

Acknowledgments

We would like to express our sincere thanks to Katsuhiko Taki, Miyuki Toku and Mari Kimura for their excellent technical assistance on this project.

References

Ebara S, Kumamoto K, Matsuura T, Mazurkiewicz JE, Rice FL (2002) Similarities and differences in the innervation of mystacial vibrissal follicle-sinus complexes in the rat and cat: a confocal microscopic study. J Comp Neurol 449:103–119

Gottschaldt KM, Vahle-Hinz C (1981) Merkel cell receptors: structure and transducer function. Science 214:183–186

Rice FL, Kinnman E, Aldskogius H, Johansson O, Arvidsson J (1993) The innervation of the mystacial pad of the rat as revealed by PGP 9.5 immunofluorescence. J Comp Neurol 337:366–385

Voltage-Dependent Calcium Channels in Merkel Cells of Hamster Oral Mucosa

Masakazu Tazaki[1], Yuki Tazaki[1], Takashi Inoue[2] and Masaki Shimono[3]

[1]Department of Physiology, Oral Health Science Center, Tokyo Dental College, 1-2-2, Masago, Mihamaku, Chiba 261-8502, Japan

[2]Department of Clinical Pathophysiology, Oral Health Science Center, Tokyo Dental College, 1-2-2, Masago, Mihamaku, Chiba 261-8502, Japan

[3]Department of Pathology, Oral Health Science Center, Tokyo Dental College, 1-2-2, Masago, Mihamaku, Chiba 261-8502, Japan

Summary

Previous studies have demonstrated the existence of voltage-dependent Ca^{2+} channels in the membrane of Merkel cells. In order to investigate which categories of Ca^{2+} channels are involved, we examined the changes in the fluorescence ratio of single isolated Merkel cells loaded with fura-2 during 334/380 nm excitation. Cells were depolarized with a solution of high KCl concentration (100 mM), known to open voltage-dependent Ca^{2+} channels. A transient increase in the fura-2 334/380 nm excitation ratio which occurred in Merkel cells was induced by 100 mM KCl. The results suggest that voltage-dependent L-type, N-type and P/Q-type Ca^{2+} channels exist in most Merkel cells. However, the number of voltage-dependent N-type Ca^{2+} channels in Merkel cells appears to be smaller or variable in comparison with the other types of Ca^{2+} channels.

Introduction

The Merkel cells-neurite complex that exists in the hamster cheek touch domes, are similar to those observed in the touch dome from primate hairy skin (Iggo and Muir 1969), thus they are considered to behave as slowly adapting type I mechanoreceptors (Tazaki and Iggo 1995). It has been hypothesized that Merkel cells from the Merkel cells-neurite complex are mechano-electric transducers (Iggo and Findlater 1984), based on the fact that voltage-dependent Ca^{2+} channels (VDCCs) exist on the Merkel cell membrane (Yamashita et al. 1992; Chan et al. 1996). However, in previous studies concerning the classification of VDCCs, there are still some aspects that need to be discussed (Yamashita et al. 1992; Chan et al. 1996). In the present study, the categories of VDCCs were investigated with specific blockers using microfluorimetric techniques in isolated single Merkel cells from hamster cheek touch domes.

Materials and Methods

In order to investigate the categories of VDCCs, isotonic highly concentrated potassium chloride solution (100 mM KCl; depolarizing stimulation) was applied to single Merkel cells using the Y-tube method and changes in fura-2AM 334/380 nm excitation ratio of single Merkel cells were observed. The excitation ratio for 100 mM KCl after applying each channel blocker was compared with the excitation ratio (control ratio) for 100 mM KCl before application of the channel blocker.

Results and Discussion

L-Type Ca^{2+} Channels

The effects of (+)-Bay K 8644 (50 μM) which is an L-type VDCCs blocker (Chan et al. 1996) were tested and are shown in Fig. 1. A transient increase in the fura-2 334/380 nm excitation ratio which occurred in Merkel cells was induced by 100 mM KCl (first black bar). Superfusion of (+)-Bay K 8644 (clear bar) showed a decrease in the fura2-AM 334/380 nm excitation ratio for the control ratio. The decrease in fura2-AM 334/380 nm excitation ratio was 39.2±6.4% (mean ± SE, range 12.3–68.1%, n=8). After washing out the (+)-Bay K 8644, a recovery of the fura2-AM 334/380 nm excitation ratio to 100 mM KCl (third black bar) was observed. These results suggested that the L-type VDCCs exist in the Merkel cells membrane.

N-Type Ca^{2+} Channels

The effects of ω-ConotoxinGVIA (1 μM) which is an N-type VDCCs blocker (Chan et al. 1996) were tested and shown in Fig. 2. Superfusion of ω-ConotoxinGVIA (clear bar) showed a decrease in the fura2-AM 334/380 nm excitation ratio for control ratio. The decrease in fura2-AM 334/380 nm excitation ratio was 13.9±0.02 % (mean±SE, range 3.0–28.4%, n=18). These results suggested that the N-type VDCCs exist in the Merkel cells membrane. However, only 18 of 25 Merkel cells (72%) examined showed a decrease in the calcium signal as mentioned above, while in 28% of the Merkel cells (7 out of 25 cells) application of ω-ConotoxinGVIA had no effect. Thus, it appears that the number of N-type VDCCs

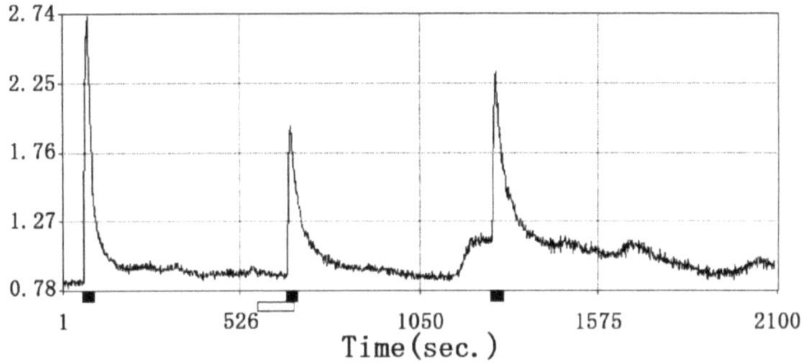

Fig. 1. Effect of (+)-Bay K 8644 (50 μM) in the intracellular influx of extracellular Ca^{2+}. *Clear bars* indicate the superfusion of (+)-Bay K 8644, *black bars* KCl

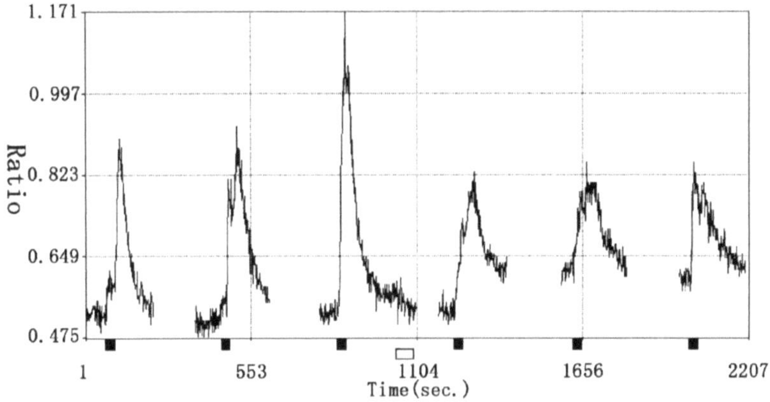

Fig. 2. Effect of ω-ConotoxinGVIA (1 μM) in the intracellular influx of extracellular Ca^{2+}. *Clear bars* indicate the superfusion of ω-ConotoxinGVIA, *black bars* KCl

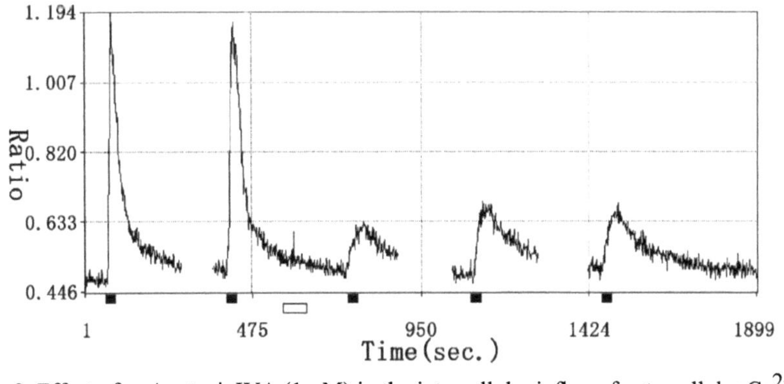

Fig. 3. Effect of ω-AgatoxinIVA (1 μM) in the intracellular influx of extracellular Ca^{2+}. *Clear bars* indicate the superfusion of ω-AgatoxinIVA, *black bars* KCl

in Merkel cells is smaller or variable in comparison with the other types of Ca^{2+} channels. Further studies regarding the N-type VDCCs will be required.

P/Q-Type Ca^{2+} Channels

The effects of ω-AgatoxinIVA (1 μM) which is a P/Q-type VDCCs blocker (Cruz et al. 1987) were tested and shown in Fig. 3. Superfusion of ω-AgatoxinIVA (clear bar) showed a decrease in the fura2-AM 334/380 nm excitation ratio for control ratio. The decrease in fura2-AM 334/380 nm excitation ratio was 27.5±0.04% (mean ± SE, range 8.7–46.7%, n=12). These results suggested that P/Q-type VDCCs exist in the Merkel cell membranes.

Although some studies have mentioned that L- or N-type VDCCs do not exist in Merkel cell membrane (Chan et al. 1996), in the present study based on the results of several investigations, it is suggested that L-type VDCCs actually exist in Merkel cell membrane (Yamashita et al. 1992). This study is also the first report regarding the existence of P/Q-type VDCCs in Merkel cell membranes.

Acknowledgments

This work was supported by a Grant-in-Aid for Scientific Research Nos. 00085771 and 40155065 to M. Shimono and M.Tazaki from the Ministry of Education, Science and Culture, Japan, and by a grant for Scientific Research No. HRC5A08 to T. Inoue from the Oral Health Science Center of Tokyo Dental College.

References

Chan E, Yung W-H, Baumann KI (1996) Cytoplasmic Ca^{2+} concentrations in intact Merkel cells of an isolated, functioning rat sinus hair preparation. Exp Brain Res 108:357–366

Cruz LJ, Johnson DS, Olivera BM (1987) Characterization of the ω-conotoxin target. Evidence for tissue-specific heterogeneity in calcium channel types. Biochemistry 26:820–824

Iggo A, Findlater GS (1984) A review of Merkel cell mechanisms. In: Hamann W, Iggo A (eds) Sensory receptor mechanisms. World Scientific, Singapore, pp 117–131

Iggo A, Muir AR (1969) The structure and function of a slowly adapting touch corpuscle in hairy skin. J Physiol 200:763–796

Tazaki M, Iggo A (1995) SAI-type mechanoreceptors in the buccal pouch of the golden hamster (*Mesocricetus auratus*). Prim Sensory Neuron 1:157–169

Yamashita Y, Akaike N, Wakamori M, Ikeda I, Ogawa H (1992) Voltage-dependent currents in isolated single Merkel cells of rats. J Physiol 450:143–162

Localization of Signal Transduction Proteins in the Merkel Cell Axon Complex

Tamiko Tachibana and Tokio Nawa

Department of Oral Anatomy, School of Dentistry, Iwate Medical University, Morioka, Japan

Summary

The function of the Merkel cell (MC) is still enigmatic. Immunohistochemically, we investigated the localization of signal transduction proteins in MC-axon complexes. Axon terminals connected to MCs in the skin and oral mucosa of the rat and monkey showed positive immunoreactivities for Gαo, Gαi-1 and Gαs, whereas Gαi-, Gαo-, Gαq-, Gαs- and Gαz-like immunoreactions were localized on MCs of both animals. Since Gαq is known to couple with group 1 metabotropic glutamate receptors and activate beta-type phospholipase C, we immunohistochemically investigated the localization of these substances in rat MCs. It was found that MCs express mGluR5-like immunoreactions. However, intensely positive immunoreactions for PLCβ4 were found to be localized in type 1 sensory nerve terminals connected to MCs, but not in MCs themselves.

Introduction

The Merkel cell (MC)-axon complex has been shown to be involved in the reception of mechanical stimuli applied to the skin and oral mucosa (Iggo and Findlater 1984; Ogawa 1996). However, the function of the MC in the complex is still controversial, though two possible functions, the mechanical transduction function and neuromodulatory function, have been proposed. We have recently shown that adult mammalian oral mucosa usually contains two subtypes of MCs, noninnervated and innervated subtypes (Tachibana et al. 1997, 1998). However, it is not known whether these two subtypes of MCs have different functions.

The localization of various kinds of potential candidates for neurotransmitters or neuromodulators, such as ATP, serotonin and various kinds of neuropeptides including opioid peptides, substance P, VIP and CGRP, in MCs has been histochemically demonstrated (Hartschuh et al. 1986; Tachibana 1995). However, little is known about the localization of the membrane receptors for these substances in the MC-axon complex. Although recent advances in gene cloning techniques have made it possible to produce antibodies for various kinds of membrane receptors, the existence of many kinds of subtypes of receptors for one kind of neurotransmitter substance makes it difficult to immunohistochemically investigate unknown membrane receptors.

Most of the neurotransmitter-like substances in MCs, for which the localization in the cells has been demonstrated, are thought to act on target cells via seven-transmembrane receptors coupled to heterotrimeric GTP-binding proteins (G proteins). The alpha subunits of the G proteins, which include various isoforms, are responsible for the GTPase activity and the mediation of intracellular signal transduction pathways (Morris and Malbon 1999). Therefore. we immunohistochemically studied the expression patterns of G protein alpha subunit isoforms in the MC-axon complex and noninnervated MCs. We found that MCs express immunohistochemical reactions for Gαi-1, Gαo, Gαq, Gαs and Gαz, irrespective of their innervation. We also found that axon terminals in the MC-axon complexes express Gαo-, Gαi- and Gαs-like immunoreactions (Tachibana et al. 2001).

Fagan and Cahusac (2001) have recently suggested on the basis of the results of their electrophysiological studies that glutamate acts as a neurotransmitter in the MC-axon complex. Although glutamate receptors are not the seven-transmembrane type, the Gαq isoform we found in MCs is known to couple to group 1 subtypes of metabotropic glutamate receptors (mGluRs) and activate beta-type phospholipase C (PLCβ; Lee et al. 1992; Pin and Duvoisin 1995). Therefore, we also immunohistochemically investigated the localization of subtypes of group 1 mGluRs and PLCβs in MCs in sinus hair follicles and the hard palate of the rat.

Immunohistochemically Determined Localization of G Protein Alpha Subunit Isoforms

Materials and Methods

Paraffin-embedded and frozen sections of whisker pads and palatine mucosae of rats and paraffin-embedded sections of the monkey lip and buccal mucosa were used for this study. Tissues were fixed in 4% paraformaldehyde dissolved in phosphate buffer (pH 7.2) or in Zamboni's fixative by intraventricular perfusion. Polyclonal antibodies to Gαo, Gαi-1, Gαi-3, Gαq, Gαs and Gαz subunits were purchased commercially and applied by immunoperoxidase and immunofluorescence techniques. To ascertain the localization of MCs, sections prepared for immunofluorescence were doubly labeled using a monoclonal antibody to cytokeratin 20 (CK20) and polyclonal antibodies to Gα subunit isoforms. The doubly labeled specimens were observed under a confocal laser scanning microscope.

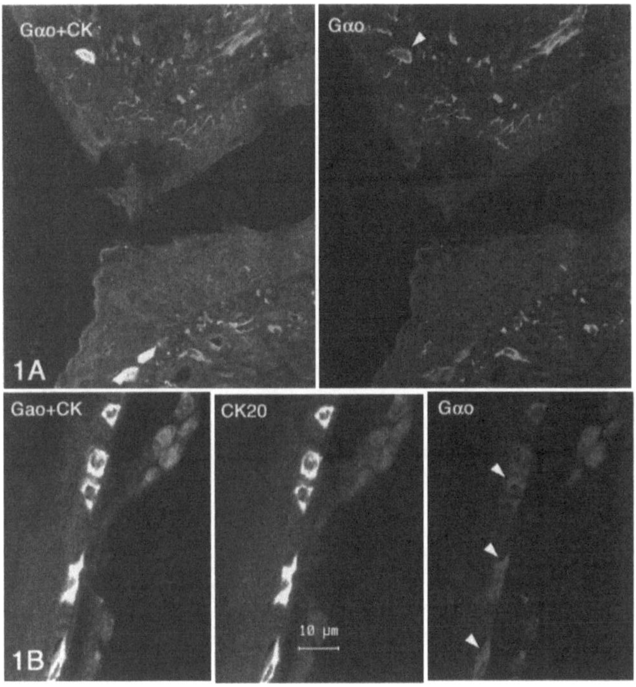

Fig. 1. Confocal laser scanning microscope images of the skin of the rat whisker pad (**A**) and the rat sinus hair follicle (**B**) double-labeled with antibodies to cytokeratin (CK) 20 and Gαo. Merkel cells in the epidermis and follicular epithelium and nerve fibers in the dermis and epidermis are labeled with anti-Gαo antibody. Note that Gαo-reactive axon terminals are in close contact with MCs (*arrowheads*)

Results

All CK 20-immunoreactive MCs in the epidermis and sinus hair follicles of the rat and monkey tissues were positively stained with anti-Gαo antibody (Fig. 1A, B). Most MCs in the rat whisker pad and the monkey tissues co-existed with nerve terminal-like structures that were also immunoreactive to anti-Gαo (Fig. 1A, B). However, many CK20-immunoreactive MCs in the rat palatine mucosa did not have apposition of nerve terminal-like structures. Nerve fibers with thin and thick calibers and their terminals in the rat and monkey tissues also showed a strong affinity to the anti-Gαo antibody (Fig. 1A, B). The intensities of all of these Gαo-like immunoreactions were diminished in control incubation using pre-adsorbed antibody solution.

Although CK 20-immunoreactive MCs and associated axon terminals in the monkey tissues were also immunoreactive for anti-Gαi-1 and -Gαs antibodies (Fig. 2), similar structures in rat tissues showed capricious reactions to these antibodies. In contrast, MCs in the rat and monkey tissues consistently showed strong positive immunoreactions for Gαq and Gαz (Fig. 3). Intensities of these reactions were diminished in control incubations using pre-adsorbed antibody solutions.

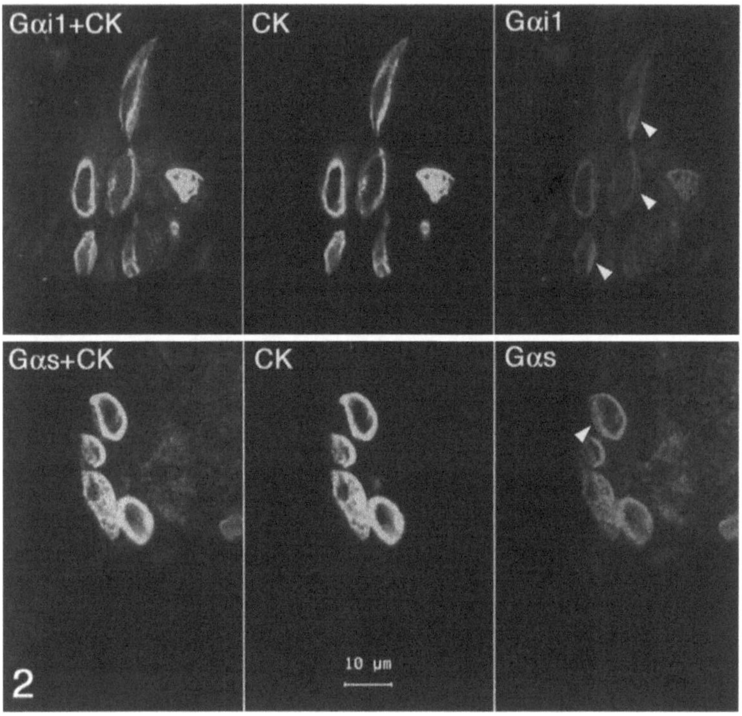

Fig. 2. Confocal laser scanning microscope images of the buccal mucosa of the monkey double-labeled with antibodies to CK20 and Gαi1 or Gαs. Axon terminals beneath MCs show positive reactions of Gαi1 and Gαs (*arrowheads*)

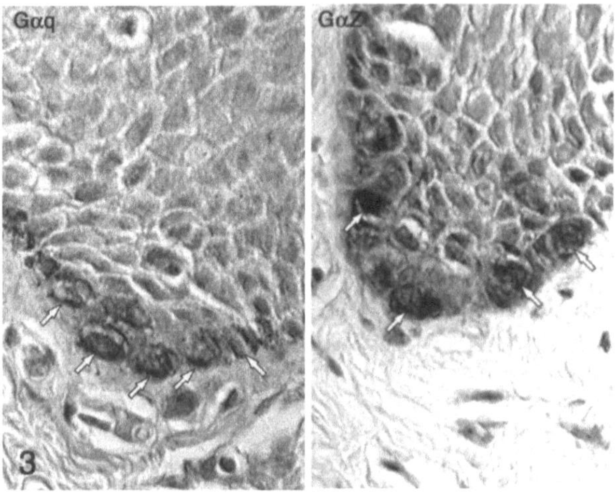

Fig. 3. Gαq- and Gαz-like immunoreactions of MCs in the buccal mucosal epithelium of the monkey stained by an immunoperoxidase technique; ×560

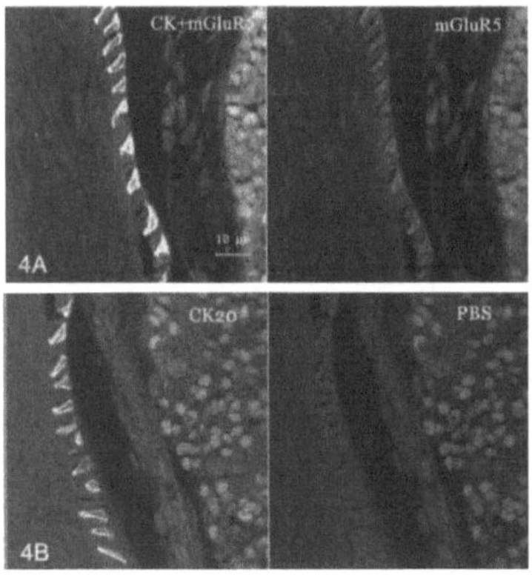

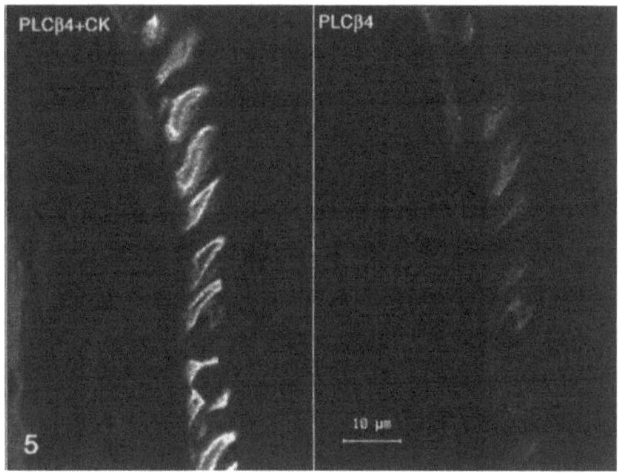

Fig. 4. Confocal laser scanning microscope images of the rat sinus hair follicle double-labeled with antibodies to CK20 (*clear white*) and mGluR5 (*dark white*) (**A**). Note that the positive mGluR5-like reactions in MCs are diminished when anti-mGluR5 was substituted with PBS (**B**)

Fig. 5. Confocal laser scanning microscope images of the rat sinus hair follicle double-labeled with antibodies to CK20 (*clear white*) and PLCβ4 (*dark white*). Positive PLCβ4-like immunoreactions are seen on nerve fibers and their terminals, but not on MCs

Immunohistochemically Determined Localization of mGluR5 and PLCβ4

Materials and Methods

Frozen and paraffin-embedded sections of paraformaldehyde-fixed whisker pads and palatine mucosa of the rat were used for this study. Polyclonal antibodies to mGluR1a, mGluR5a, PLCβ1, PLCβ2, PLCβ3 and PLCβ4 were purchased commercially and applied to the above sections by immunoperoxidase and immunofluorescence techniques.

Results

Of the two kinds of antibodies to group 1 mGluR, only anti-mGluR5a was reactive for cells in the outermost layer of the epithelium at the neck portion of sinus hair follicles and some cells distributed in the basal layer of the palatine epithelium. Double labeling confocal laser scanning microscopy revealed that the mGluR5-like immunoreactions occurred in CK 20-immunoreactive MCs (Fig. 4A). Although MCs in sections incubated in PBS instead of anti-mGluR5 antibody solution showed negative immunoreactions (Fig. 4B), we did not perform control incubations using pre-adsorbed antibodies because it was difficult to obtain an antigen protein.

Among the four antibodies to PLCβ isozymes used in this study, only anti-PLCβ4 antibody was reactive for the outermost layer of sinus hair follicular epithelium. However, unexpectedly, a double labeling immunofluorescence technique revealed that the PLCβ4-like immunoreactions occurred in nerve fibers and their terminals connected to MCs, but not in MCs themselves (Fig. 5). Although we did not ascertain the reliability of these reactions using pre-adsorbed antibodies, cerebellar Purkinje cells of the rat used for a positive control were also positively stained with the anti-PLCβ4 antibody.

Discussion

Two possible functions of MCs in cutaneous type I mechano-sensory receptors, i.e., MC-axon complexes, have been suggested. One speculation is the idea that MCs act as mechanical transducers (Iggo and Findlater 1984; Ogawa 1996) and the other is that MCs act as modulators or trophic supporters for nerve terminals (Mills and Diamond 1995). Although there have been various physiological findings supporting the former speculation (Akaike and Yamashita 1995; Tazaki and Iggo 1995; Baumann et al. 1996; Chan et al. 1996), little is known about the signal transduction pathways in the receptor.

The results of our immunohistochemical studies indicate the possible existence of Gαo, Gαi-1 and Gαs in the axon terminals of type I mechanoreceptors, though the presence of the latter two molecules in the rat is not clear. Of these alpha subunits, Gαo and Gαi are known to suppressively regulate cellular actions via reduction of adenylyl cyclase activity or closure of Ca^{2+} channels (Morris and Malbon 1999). Since G proteins containing these kinds of alpha subunit isoforms are known to couple to opioid and serotonin receptors (Morris and Malbon 1999), it is possible that these substances so far identified in the MC act as neuromodulators. However, the immunohistochemical expression of Gαs-like reactions in axon terminals of type I mechanoreceptors of the monkey suggests that MCs also act as mechanical transducers, because the action of the Gαs isoform is the opposite to that of Gαo and Gαi (Morris and Malbon 1999).

In our immunohistochemical study, MCs themselves also expressed immunoreactions for five kinds of alpha subunit isoforms of G proteins: Gαo, Gαi1, Gαq, Gαs and Gαz. The Gαo, Gαi and Gαz subunit isoforms belong to the Gi family of proteins, which mainly suppress cellular actions, whereas the Gαs isoform belongs to the Gs family of proteins, which have actions opposite to those of Gi family proteins, as mentioned above. The presence of many kinds of Gα-subunit isoforms in MCs suggests that the metabolism and actions of these cells are regulated by neurotransmitters or hormones. The contents of MC granules may act on MCs themselves by autocrine mechanisms.

The Gαq subunit couples to mGluRs and regulates the intracellular Ca^{2+} concentration via the activation of phosphoinositide-specific PCLβ (Lee et al. 1992; Pin and Duvoisin 1995). As mentioned above, Fagan and Cahusac (2001) have recently reported that glutamate probably acts as a neurotransmitter in type I mechanoreceptors in the skin. We therefore speculated that glutamate acts not only on axon terminals, but also on MCs themselves and regulates the secretion of MC granules during or after the excitation of type I mechanoreceptors by a feedback mechanism. Actually, we were able to determine that mGluR5-like immunoreactions were localized in MCs in the rat sinus follicles. However, these cells were negative for all antibodies to PLCβ isozymes that we used, though further studies using other antibodies are needed to reach a conclusion.

Axons and their terminals innervating MCs in the rat hair follicles unexpectedly showed strong positive immunoreactions for PCLβ4 in our study. The positive immunoreactions seem to reflect the presence of the isozyme, because the antibody to PLCβ4 that we used was also reactive for cerebellar Purkinje cells, which are known to have this isozyme (Min et al. 2000). The significance of the existence of PLCβ4 in the axon terminals is obscure at the present. However, it is known that PLCβ isozymes are also activated by low-molecular weight G proteins and βγ subunit complexes of Gi and Go proteins (Lee et al. 1992; Pin and Duvoisin 1995). Further studies are needed to determine the significance of this enzyme in the axon terminals of type I mechanoreceptors.

Previously, we reported that mammalian oral mucosae contain two peculiar subtypes of MCs, innervated and noninnervated types (Tachibana et al. 1997, 1998). These two subtypes were found to be identical not only in the expression pattern of G protein alpha subunit isoforms, but also in the expression of the mGluR5. Gα-subunits found in MCs in our study may be related to the fundamental metabolism or the survival of both types of MCs. However, further study is needed to determine the G proteins that are specific to individual types of MCs.

We believe that clarification of signal transduction pathways as well as membrane receptors in MCs and related axon terminals is required for an understanding of the function of MCs. Further studies in this field are awaited.

Acknowledgements

This work was in part supported by grants-in-aid for scientific research from the Ministry of Education, Science and Culture of Japan (No. 14570024, Open Research Project).

References

Akaike N, Yamashita Y (1995) The physiology and ion channels of the Merkel cell. In: Heschele J, Scherubl H (eds) The electrophysiology of neuroendocrine cells. CRC Press, Boca Raton, pp 317–328

Baumann KI, Chan E, Halata Z, Senok SS, Yung W-H (1996) An isolated rat vibrissal preparation with stable responses of slowly adapting mechanoreceptors. Neurosci Lett 213:1–4

Chan E, Yung W-H, Baumann KI (1996) Cytoplasmic Ca^{2+} concentrations in intact Merkel cells of an isolated, functioning rat sinus hair preparation. Exp Brain Res 108:357–366

Fagan BM, Cahusac PMB (2001) Evidences for glutamate receptor-mediated transmission at mechanoreceptors in the skin. Neuroreport 12:341–347

Hartschuh W, Weihe E, Reinecke M (1986) The Merkel cell. In: Breiter-Hahn AGJ, Matoltsy AG, Rechards KS (eds) Biology of integument 2. Springer, Berlin Heidelberg New York, pp 606–620

Iggo A, Findlater GS (1984) A review of Merkel cell mechanisms. In: Hamann W, Iggo A (eds) Sensory receptor mechanisms. World Scientific Publications, Singapore, pp 117–131

Lee CH, Park D, Wu D, Rhee SG, Simon MI (1992) Members of the Gqα subunit gene family activate phospholipase Cβ isozymes. J Biol Chem 267:16044–16047

Mills LR, Diamond J (1995) Merkel cells are not the mechanosensory transducers in the touch dome of the rat. J Neurocytol 24:117–134

Min DS, Kim MJ, Jeong HK, Lee YH, Kim H, Shin H-S, Ryu SH, Suh P-G (2000) Immunological characterization of 130-Kda phospholipase Cβ4 isozyme in rat cerebellar Purkinje cells. Neurosci Lett 292:9–12

Morris AJ, Malbon CC (1999) Physiological regulation of G protein-linked signaling. Physiol Rev 79:1373–1430

Ogawa H (1996) The Merkel cells as a possible mechanoreceptor cells. Progr Neurobiol 49:317–334

Pin JP, Duvoisin R (1995) The metabotropic glutamate receptors: structure and functions. Neuropharmacology 34:1–26

Tachibana T (1995) The Merkel cell: recent findings and unresolved problems. Arch Histol Cytol 58:376–396

Tachibana T, Yamamoto H, Takahashi N, Kamegai T, Shibanai S, Iseki H, Nawa T (1997) Polymorphism of Merkel cells in the rodent palatine mucosa: immunohistochemical and ultrastructural studies. Arch Histol Cytol 60:379–389

Tachibana T, Kamegai T, Takahashi N, Nawa T (1998) Evidence for polymorphism of Merkel cells in the adult human oral mucosa. Arch Histol Cytol 61:115–124

Tachibana T, Endoh M, Nawa T (2001) Immunohistochemical expression of G protein α-subunit isoforms in rat and monkey Merkel cell-neurite complexes. Histchem Cell Biol 116:205–213

Tazaki M, Iggo A (1995) SAI-type mechanoreceptors in the buccal pouch of the golden hamster (*Mesoricetus auratus*). Prim Sens Neur 1:157–169

Are Glutamate Receptors Involved in Transmission at the Junction Between Merkel Cell and Nerve Terminal?

Peter Cahusac

Department of Psychology, University of Stirling, Stirling FK9 4LA, Scotland, UK

Summary

This review briefly surveys the histological and pharmacological evidence for a neurotransmitter in the Merkel cell-to-nerve ending junction. A recent pharmacological study obtained some clear evidence that glutamate receptors were involved, implicating some kind of amino acid transmitter. However, further pharmacological work, aimed at characterizing the specific glutamate receptor, produced results that were difficult to interpret. Thus, it is possible that some kind of unconventional NMDA glutamate receptor is involved. Alternatively, the results question whether glutamate receptors are directly involved at this junction.

Introduction

Merkel cells have an unusually close association with primary afferent nerve endings in the skin. This association may serve one or more purposes (Tachibana 1995): for example, they may play a key support or regulatory role in the basal epidermis, and may provide targets for developing or regenerating nerves. They may also act as mechanoreceptors – conveying information to those closely apposed primary afferent terminals. Merkel himself appears to have favoured this last possibility since he referred to his cells as 'Tastzellen' (touch cells). Much later, histological studies showed that the junction between the Merkel cell and nerve ending resembled a synapse, and that the Merkel cell contained dense core vesicles – presumably containing a chemical transmitter necessary to traverse the junction. Iggo and Muir (1969), studying the so-called touch domes of hairy skin, first revealed a direct correlation between Merkel cells and electrical activity recorded from slowly adapting type I single nerve fibres. Perhaps the clearest scenario describing a neurotransmitter from Merkel cell-to-nerve ending is provided in a review many years later by Iggo and Findlater (1984). A more thorough and updated review of the evidence was given by Ogawa (1996). Since then further work has continued to support the hypothesis that Merkel cells are mechanoreceptors (Christie et al. 1996; Senok and Baumann 1997; Tazaki and Suzuki 1998; Halata et al. 1999).

Histological Evidence

The idea that a neurotransmitter was involved spurred on many histological studies. These showed variously that ATP, 5-HT and a range of neuropeptides, were located in Merkel cells. However, there were considerable interspecies differences (particularly for the neuropeptides). For example, met-enkephalin was found in the rat, but not cat, and vice versa for vasoactive intestinal peptide (VIP; Tachibana 1995). According to Occam's razor, one would expect only one principal neurotransmitter to be involved across all or most species. It is possible that some of these histologically localized substances modulate Merkel cell activity rather than transmit fast excitation, and so another recent approach was to localize G protein subunits necessary for modulation, without identifying the specific modulator. One study found Merkel cells and associated nerve terminals labelled for at least two different G proteins (Tachibana et al. 2001).

Recently, it has been found that sensory axons in the skin were labelled for glutamate receptors (Kinkelin et al. 2000), although it was not possible to identify the class of axons or whether they were associated with Merkel cells. Another study, this time of keratinocytes, revealed clear evidence that glutamate was involved in signalling between cells (Genever et al. 1999). Close inspection of illustrations from that work suggested that at least some of the immuno-labelled cells of the basal epidermis were Merkel cells. This encouraged further histological work, which is now reported in another chapter of this volume (Senok et al., this Vol.).

Histological studies can provide evidence that is *necessary* to implicate a particular neurotransmitter. Only functional (e.g. pharmacological) studies are capable of providing confirmatory evidence that is *sufficient.*

Pharmacological Approaches

A number of studies used a direct pharmacological approach to try to block chemical neurotransmission. Iggo and Muir (1969) tried systemic depletion of vesicular transmitter stores using reserpine. There were no changes in neural responses to mechanical stimulation, nor depletion of the vesicles, following this treatment. Another approach was to apply selective receptor agonists or antagonists to skin containing Merkel cells in vivo, either topically (Smith and Creech 1967), by close-arterial injection (Fjällbrant and Iggo 1961), or by intravenous infusion (Gottschaldt and Vahle-Hinz 1982). Alternatively, drugs were applied in vitro by superfusion of a patch of skin in an isolated tissue bath (Baumann and Chan

1993) or by intra-arterial injection of an in vitro tissue preparation (Martin et al. 1990). In each of these methods neural responses to mechanical stimulation were recorded. Together, the studies assessed the following neurotransmitters: acetylcholine, 5-HT, ATP, histamine, adrenaline, noradrenaline, dopamine, GABA, glycine, met-enkephalin, dynorphin, among others. Of the compounds tested, 5-HT, histamine, ATP, acetylcholine and nicotine had effects – some depressant, some excitatory – though there was some disagreement between different reports. Excitatory amino acid (EAA) receptor antagonists were never tested.

Excitatory Amino Acids and Receptors

There are two reasons why an EAA is a good transmitter candidate. Firstly, glutamate receptors, at which these transmitters act, account for most fast chemical neurotransmission in the central nervous system. Secondly, in other sensory systems, there is increasing evidence that receptor cells make glutamatergic synapses with associated primary afferent nerve terminals. For example, in the vertebrate retina, photoreceptors transmit to bipolar cells through glutamate receptors (Brandstatter and Hack 2001). In the cochlea, the junction between the mechanosensitive inner hair cells and type I auditory primary afferents is glutamatergic (Puel 1995). In the latter case, the morphological and functional similarities between inner hair cells and Merkel cells was previously noted by Iggo and Findlater (1984), and gives added support to the idea that an EAA might be involved at the Merkel cell junction. It was in the light of this speculation that a direct test of whether glutamate receptors mediated transmission from Merkel cells was attempted (Fagan and Cahusac 2001; see below).

There are a number of candidates for an EAA transmitter: primarily l-glutamate and l-aspartate, but also l-cysteate, l-cysteine sulfinate, l-homocysteate, l-homocysteine sulfinate, and quinolinic acid among others (Watkins and Evans 1981; Stone and Burton 1988). These amino acids act at two classes of glutamate receptor: ionotropic and metabotropic. Ionotropic glutamate receptors are ligand-gated channels and tend to mediate fast transmission. There are three main types of ionotropic receptor: NMDA, AMPA and kainate. In contrast, the class of metabotropic glutamate receptors are coupled to G proteins, and their activation is usually responsible for slower transmission or modulation of transmission. There are three groups of receptor types within this class. It is of interest that group I metabotropic glutamate receptors were found on Merkel cells and associated nerve terminals (Tachibana and Nawa, this Vol.).

Testing Glutamate Receptor Antagonists

In experiments to examine the possibility of an EAA transmitter, the isolated rat sinus hair preparation (Baumann et al. 1996) was used. Various concentrations

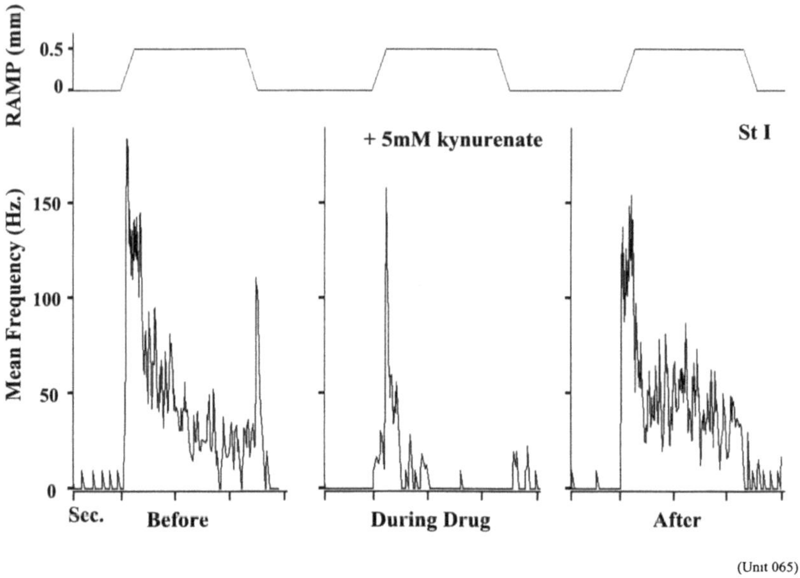

Fig. 1. Depressant effect of 5 mM kynurenate on St I unit responses to a mechanical ramp stimulus

of kynurenate, a broad spectrum ionotropic glutamate receptor antagonist, were tested against responses of sinus hair type I (St I) units evoked by mechanical displacement of the hair. Kynurenate 1–10 mM depressed St I unit responses in a concentration-dependent and reversible manner (Fagan and Cahusac 2001). The static phase of the response (that to maintained whisker deflection) was particularly susceptible (see Fig. 1). Responses of St II units were unaffected by kynurenate in this concentration range. St II units are believed to be associated with other receptors comprised of specialized nerve endings, such as lanceolate mechanoreceptors, which do not involve synapse-like junctions. It is conceivable that the lower concentration needed to depress St I versus St II was due to differential access (due to physical or chemical barriers) to the mechanoreceptors, although this seems unlikely since other drugs were shown to differentially affect St II responses compared with St I responses. For example caffeine, at the same concentration, was shown to simultaneously depress St II responses and enhance St I responses (Senok and Baumann 1997). The more pronounced effect of kynurenate on St I unit static phase compared with the dynamic phase was in agreement with the two-receptor-sites hypothesis (Yamashita and Ogawa 1991). This hypothesis proposes that type I responses are due to activation of mechanically gated channels in both the nerve endings and in Merkel cells. Stated briefly, the activation of the nerve endings is responsible for the high frequency burst of the dynamic response phase (the initial response to the moving mechanical stimulus), and activation of Merkel cells is responsible for the lower frequency static response.

The positive result with kynurenate appeared to establish that the junction between the Merkel cell and nerve terminal was glutamatergic. All that was now required was to identify the particular type(s) of receptor. A range of selective glutamate receptor antagonists was therefore tested. Disappointingly, the most potent and selective NMDA and non-NMDA receptor antagonists had little effect on St I responses. However, MK 801, an uncompetitive NMDA receptor blocker, caused clear and long-lasting depression of responses and spontaneous activity (Senok et al. 2001). To complicate matters, there were some peculiarities of MK 801's effect. First, the effect had a rather narrow effective concentration range (a steep concentration-response slope) with 100 μM as a very approximate IC50 (concentration at which the response is reduced by 50% from control). A complete block of all activity was usually obtained with 300 μM. This meant that many tests gave an all-or-nothing result. Second, both stereoisomers of MK 801 were equipotent. This was curious since (–)-MK 801 is reckoned to be tenfold less potent than (+)-MK 801 in its action at NMDA receptors. However, there was some discrimination between St I and St II units, but not much. St II units appeared to be less susceptible to both isomers of MK 801, showing little or no effect at 100 μM and clear depression at 300 μM. A very similar picture of results was subsequently obtained with another uncompetitive NMDA receptor antagonist, ketamine. Finally, Group I and II metabotropic glutamate receptor antagonists were tested: neither had clear or reproducible effects on responses.

Interpreting the Results with Antagonists

What is to be made from this set of results? We return first to the depressant effect of kynurenate on St I responses. In other studies, it was found that kynurenate most potently antagonises the strychnine-insensitive glycine site of the NMDA receptor, this occurring at 10 μM. At higher concentrations, 0.1–1 mM, it antagonises AMPA and kainate receptors (Stone 1993). In our study, the depression of St I responses occurred only at relatively high concentrations of kynurenate (IC50 of about 3.5 mM), suggesting a possible action at non-NMDA (AMPA or kainate) receptors. However, attempts to block responses with selective NMDA and non-NMDA receptor antagonists failed. It has often been assumed that kynurenate was a relatively "clean" drug, in that it appeared to be a selective (and competitive) receptor antagonist at ionotropic and metabotropic glutamate receptors (Stone 1993). Recent work has questioned kynurenate's selectivity by showing that it had a noncompetitive inhibitory action at nicotinic acetylcholine receptors at concentrations of less than 10 μM (Hilmas et al. 2001). A direct comparison showed that kynurenate more effectively inhibited nicotinic receptors than NMDA receptors at the surprisingly low concentration range of 0.1–1 μM.

It is uncertain at what site many drugs or receptor antagonists (kynurenate, MK 801, ketamine) are acting. As well as the obvious and intended post-junctional site, they may be acting at extra-junctional sites on the nerve endings or directly on Merkel cells. Further tests with other selective receptor antagonists will be necessary. The interpretation of drug effects would be significantly aided if direct electrical recordings could be made from single Merkel cells. Unfortunately, isolated Merkel cells (Yamashita et al. 1992; Ogawa 1996), which would allow such recording, have been found to be unresponsive to mechanical stimulation, perhaps due to enzymatic treatment. It is possible that cultured Merkel and primary afferent ganglion cells (Vos et al. 1991) would offer a way to record from both pre- and post-synaptic cells.

Conclusion

Initial work with kynurenate provided some encouraging support for the hypothesis that glutamate receptors mediate transmission at the junction from Merkel cell to nerve terminals. However, this was contradicted by subsequent work with more selective receptor antagonists. Only one class of NMDA receptor antagonists (here represented by MK 801 and ketamine) had clear-cut depressant effects on St I activity. This class of receptor antagonists are specifically known as uncompetitive blockers of an open-channel associated with the NMDA receptor. The idiosyncratic effects of these antagonists on St I activity described above, suggests that a conventional form of the NMDA receptor (NR1 with NR2A-D subunits) is not involved on post-junctional nerve terminals. There are a number of possible sites of action for these compounds, such as a direct action on Merkel cells or at extra-junctional sites. As it stands, the data concerning the effects of EAA receptor antagonists cast strong doubts on the involvement of an EAA as the transmitter.

At a previous Merkel Cell symposium, held in Tokyo in 1999, Ogawa (2000) concluded his discussion of Merkel cells as possible mechanoreceptors by suggesting that the transmitter will be identified using specific antagonists, and that ". . . such evidence will be obtained in the next 5 years." Now, in 2003, this kind of approach may yet be on target.

Acknowledgements

The work reviewed here was supported by the Wellcome Trust (ref: 061023) and the University of Stirling. I am grateful to Dr Solomon Senok for discussions and comments on the manuscript.

References

Baumann KI, Chan E (1993) Met-enkephalin does not function as a neurotransmitter in slowly adapting type I cutaneous mechanoreceptors in an isolated rat skin-nerve preparation. J Physiol 459:207P

Baumann KI, Chan E, Halata Z, Senok SS, Yung WH (1996) An isolated rat vibrissal preparation with stable responses of slowly adapting mechanoreceptors. Neurosci Lett 213:1–4

Brandstatter JH, Hack I (2001) Localization of glutamate receptors at a complex synapse: the mammalian photoreceptor synapse. Cell Tissue Res 303:1–14

Christie KN, Thomson C, Hopwood D, Harper AA, Ogden GR (1996) An immunocytochemical study of the carbonic anhydrase i isoenzyme in human oral Merkel cells. Arch Oral Biol 41:901–904

Fagan BM, Cahusac PMB (2001) Evidence for glutamate receptor-mediated transmission at mechanoreceptors in the skin. Neuroreport 12:341–347

Fjällbrant N, Iggo A (1961) The effect of histamine, 5-hydroxytryptamine and acetylcholine on cutaneous afferent fibres. J Physiol 156:578–590

Genever PG, Maxfield SJ, Kennovin GD, Maltman J, Bowgen CJ, Raxworthy MJ, Skerry TM (1999) Evidence for a novel glutamate-mediated signaling pathway in keratinocytes. J Invest Dermatol 112:337–342

Gottschaldt KM, Vahle-Hinz C (1982) Evidence against transmitter function of met-enkephalin and chemosynaptic impulse generation in "Merkel cell" mechanoreceptors. Exp Brain Res 45:459–463

Halata Z, Cooper BY, Baumann KI, Schwegmann C, Friedman RM (1999) Sensory nerve endings in the hard palate and papilla incisiva of the goat. Exp Brain Res 129:218–228

Hilmas C, Pereira EFR, Alkondon M, Rassoulpour A, Schwarcs R, Albuquerque EX (2001) The brain metabolite kynurenic acid inhibits [alpha]7 nicotinic receptor activity and increases non-[alpha]7 nicotinic receptor expression: physiopathological implications. J Neurosci 21:7463–7473

Iggo A, Muir AR (1969) The structure and function of a slowly adapting touch corpuscle in hairy skin. J Physiol 200:763–796

Iggo A, Findlater GS (1984) A review of Merkel cell mechanisms. In: Hamann W, Iggo A (eds) Sensory receptor mechanisms. World Scientific, Singapore, pp 117–131

Kinkelin I, Brockner EB, Koltzenburg M, Carlton SM (2000) Localization of ionotropic glutamate receptors in peripheral axons of human skin. Neurosci Lett 283:149–152

Martin HA, Tuckett RP, English K (1990) 5-HT and ATP modulate the activity of cutaneous type I mechanoreceptors in the rat: an "in vitro" study. Soc Neurosci Abstr 16:703

Ogawa H (1996) The Merkel cell as a possible mechanoreceptor cell. Progr Neurobiol 49:317–334

Ogawa H (2000) Merkel cells: possible mechanoreceptors. In: Suzuki H, Ono T (eds) Merkel cells, Merkel cell carcinoma and neurobiology of the skin. Elsevier, Amsterdam, pp 83–87

Puel J-L (1995) Chemical synaptic transmission in the cochlea. Progr Neurobiol 47:449–476

Senok SS, Baumann KI (1997) Functional evidence for calcium-induced calcium release in isolated rat vibrissal Merkel cell mechanoreceptors. J Physiol 500:29–37

Senok SS, Baumann KI, Cahusac PMB (2001) MK-801 depression of vibrissal slowly adapting mechanoreceptor responses. J Physiol 536P:38
Smith KR, Creech BJ (1967) Effect of pharmacological agents on the physiological responses of hair discs. Exp Neurol 19:477–482
Stone TW (1993) Neuropharmacology of quinolinic and kynurenic acids. Pharmacol Rev 45:309–379
Stone TW, Burton NR (1988) NMDA receptors and ligands in the vertebrate CNS. Progr Neurobiol 30:333–368
Tachibana T (1995) The Merkel cell: recent findings and unresolved problems. Arch Histol Cytol 58:379–396
Tachibana T, Endoh M, Nawa T (2001) Immunohistochemical expression of G protein alpha-subunit isoforms in rat and monkey Merkel cell-neurite complexes. Histochem Cell Biol 116:205–213
Tazaki M, Suzuki T (1998) Calcium inflow of hamster Merkel cells in response to hyposmotic stimulation indicate a stretch activated ion channel. Neurosci Lett 243:69–72
Vos P, Stark F, Pittman RN (1991) Merkel cells in vitro: production of nerve growth factor and selective interactions with sensory neurons. Dev Biol 144:281–300
Watkins JC, Evans RH (1981) Excitatory amino acid transmitters. Ann Rev Pharmacol Toxicol 21:165–204
Yamashita Y, Ogawa H (1991) Slowly adapting cutaneous mechanoreceptor afferent units associated with Merkel cells in frogs and effects of direct currents. Somatosens Mot Res 8:87–95
Yamashita Y, Akaike N, Wakamori M, Ikeda I, Ogawa H (1992) Voltage-dependent currents in isolated single Merkel cells of rats. J Physiol 450:143–162

Glutamate Receptor-Like Immunoreactivity in Rat Vibrissal Merkel Cells

S.S. Senok[1], P.G. Genever[2], P.M.B. Cahusac[3], and K.I. Baumann[4]

[1]Department of Physiology, College of Medicine and Medical Sciences, Arabian Gulf University, P.O. Box 22979, Manama, Bahrain

[2]Department of Biology, University of York, York, UK

[3]Department of Psychology, University of Stirling, Stirling FK9 4LA, Scotland, UK

[4]Department of Functional Anatomy, University of Hamburg, Hamburg, Germany

Summary

Recent functional data has raised the possibility that an excitatory amino acid, such as glutamate, may act as neurotransmitter in vibrissal type I slowly adapting (St I) mechanoreceptors. It is supposed that Merkel cells respond to mechanical stimulation by releasing a neurotransmitter which acts on the nerve terminals in contact with the Merkel cells. Therefore, we sought to establish whether or not the appropriate signalling molecules for glutamate neurotransmission are to be found in the vibrissa. Using frozen sections from vibrissa isolated from adult Wistar rats, we stained for NMDA and AMPA glutamate receptors and the glutamate transporter GLT-1. Distinct patterns of expression of NMDA-type receptors were found in the vibrissal sections. There was intense immunoreactivity to the NR1 antibody in the layer of cells in the outer root sheath, adjacent to the glassy membrane. However, NR2A/B staining was restricted to a subpopulation of cells adjacent to the glassy membrane, but only above the level of the ring wulst and the innermost cells of the rete ridge region – suggesting that the NR2A/B stained cells are Merkel cells. GLT-1 immunoreactivity was largely restricted to the epithelial cells lining the thick sinus body capsule enclosing the blood sinuses. There was no remarkable staining with the AMPA-type GluR1–3 antibodies. The data suggest the presence of glutamate signaling in the outer root sheath cells, including Merkel cells. However, it remains to be seen if this is related to mechanosensory function in Merkel cells or whether it is part of the already described glutamate signaling in keratinocytes.

Introduction

The anatomical association between Merkel cells and nerve terminals in the skin and appendages across many species is suggestive of a synaptic relationship. Although functional and other data (Yamashita and Ogawa 1991; Chan et al. 1996; Senok and Baumann 1997) continue to support the notion that the Merkel cell is a mechanosensory transducer which responds to mechanical stimulation by releasing a transmitter, the identity of the transmitter has remained elusive (see reviews in this volume; Ogawa (1996). Recent functional data has raised the possibility that glutamate may be a chemical transmitter at vibrissal Merkel cell touch receptors (Fagan and Cahusac 2001; Senok et al. 2001). It was hypothesized that if glutamate served a transmitter function between Merkel cells and nerve terminals, the appropriate glutamate signalling molecules should be present wherever Merkel cell-neurite complexes are found.

Vibrissae were used because they have a high concentration of Merkel cells and nerve terminals in a compact, well-defined location and the fact that the functional work (Fagan and Cahusac 2001; Senok et al. 2001) used isolated vibrissae.

Glutamate is the major excitatory neurotransmitter in the brain. It exerts its effects through ionotropic (fast synaptic transmission) and metabotropic receptors (modulation of pre- and post-synaptic activity). The ionotropic receptors are subdivided into NMDA and non-NMDA (comprising AMPA and Kainate receptors), based on their sensitivity to ligands. The NMDA receptor is made up of a multimeric assembly of four or five subunits, comprising NR1 and one or more NR2 (A–D) subunits. Expression of both types of subunit is required to form functional channels. The non-NMDA receptors are also made up of subunits – GluR1–4 for AMPA, and GluR5–7 and KA1–2 for Kainate. The transporters EAAC1, GLT-1 and GLAST transport glutamate away from the synaptic cleft. Therefore, we immunostained the vibrissae for the following signalling molecules: NR1, NR2 A/B, GluR1–3 and GLT-1.

Materials and Methods

Antibodies

All antibodies were obtained from Chemicon (Harrow, UK), unless otherwise stated. Rabbit polyclonal anti-NR1, NR2A/B, GluR1, and GluR2/3 (UBI, New York, USA) were used. Anti-GLT-1 polyclonal antibody was provided by Dr. Jeffrey Rothstein (Johns Hopkins, USA).

Immunolocalisation of Glutamate Receptors and Transporters in Rat Vibrissae

Vibrissae were dissected from adult Wistar rats as previously described (Senok et al. 1996). Isolated vibrissa were dipped in 10% polyvinyl alcohol (PVA; Sigma,

Poole, UK), immediately frozen in chilled isopentane (–70 °C) and mounted in 10% PVA on brass chucks. Sections (5–7 μm thickness) were cut using a Bright cryostat (Bright Instrument Co., Huntington, UK), collected on polysine slides (BDH) and stored at –35 °C until use.

The sections were fixed in 4% paraformaldehyde for 5 min and endogenous peroxidase activity depleted with 3% hydrogen peroxide (Sigma) for 30 min. A further pre-incubation was performed with 10% normal goat serum (Vector Laboratories) for 30 min to block non-specific antibody binding. Sections were incubated for 30 min with primary polyclonal antibody (individual antibodies were titrated on each tissue to determine optimal concentration and a range of 0.5–1 μg/ml was used) followed by biotinylated goat anti-rabbit secondary antibody (Vector Laboratories; 1:200 dilution) for 15 min and avidin-biotinylated-peroxidase reagent (ABC Elite, Vector Laboratories, 1:50 dilution) for 20 min. Peroxidase activity was disclosed with 0.5 mg/ml 3, 3′-diaminobenzidine (Sigma) and 0.3% hydrogen peroxide as substrate. All dilutions were made up in phosphate-buffered saline (PBS), pH 7.4 (GLT-1 antibodies were diluted in PBS containing 0.1% Triton X-100) and incubations were performed at room temperature with three PBS washes between each incubation. Negative controls received the same concentration of normal rabbit IgGs (Vector Laboratories) in place of primary antibody. Sections were counterstained with haematoxylin prior to mounting in glycerol/PBS.

Results

NMDA Receptor-Like Immunoreactivity

NR1. There was uniform immunostaining of the monolayer of cells of the external root sheath both above and below the ring wulst (Fig. 1e,f).

NR2 A/B. There was staining of a subset of the outer root sheath cells above the ring wulst (Fig. 1a). A layer of cells in the rete ridge (Fig. 1b) was also clearly immunopositive for NR2 A/B, suggesting that these are Merkel cells. A circumscribed band of non-cellular profiles that appear to be embedded in the glassy membrane was seen about the middle third of the area distal to the ring wulst, in the area to the cavernous sinus (Fig. 1c,d).

GluR1, GluR2 and Glur3 did not show any clear staining in the hair follicle (Fig. 2a–c).

The anti-GLT1 transporter antibody clearly stained a layer of apparent endothelial cells lining the capsule of the blood sinus (Fig. 2d).

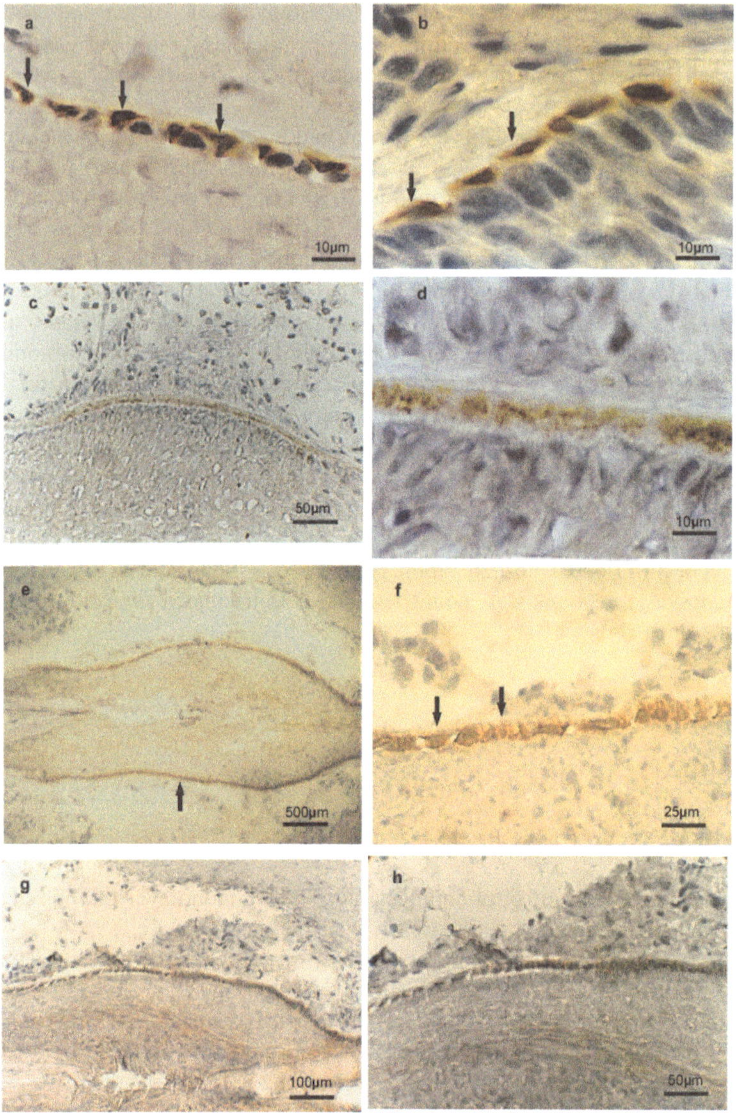

Fig. 1. NR2 A/B immunoreactivity (*brown* reaction product) of a subset of external root sheath cells above the ring wulst (**a**, *arrows*) and the group of cells in the rete ridge collar known to be Merkel cells (**b**, *arrows*). **c, d** Band of intensely stained profiles in contact with the glassy membrane at the level of the cavernous sinus. Unlike the NR2 A/B staining, NR1 staining was intense and uniform in the external root sheath cells above and below the ring wulst (**e**, *arrow*; and **f**, higher power showing the stained cells). Negative control slides show no staining of external root sheath cells (**g**, **h**)

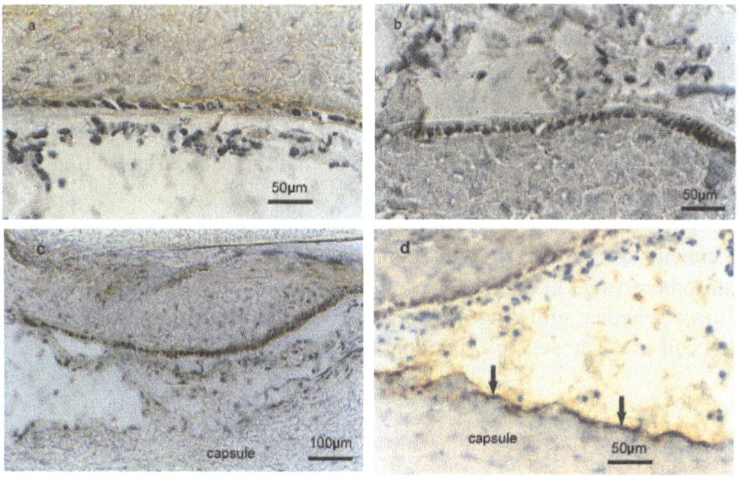

Fig. 2. AMPA-type GluR1, 2 and 3 showed no clear immunoreactivity in the external root sheath cells (**a**, **b**, **c**). GLT-1 is seen in the endothelial lining of the sinus capsule (**d**, *arrows*; cf. **c** and Fig. 1g). Elements of the mesenchymal tissue in the blood sinus also appear to be positive for GLT-1

Discussion

The positive staining for NR1, NR2 A/B and GLT-1 in the outer root sheath and the rete ridge collar of the sinus hair follicle suggests that the signalling molecules for glutamate transmission are present in the appropriate location of the vibrissa. The differential distribution of the NMDA receptor subunits in the external root sheath cells might reflect functional differences between the cells.

If glutamate is the transmitter, the ionotropic receptors ought to be on the nerve terminals rather than the Merkel cells (the metabotropic receptors would be expected on the Merkel cells as reported by Tachibana and Nawa (this Vol.). While it is clear that the NR1 staining is in cells, we are unable to categorically say whether the NR2 A/B is on the Merkel cells or the expanded nerve terminals. Further work is required to clarify this. In any event, the processes involving (non-sensory) glutamate signalling described in the skin (Genever et al. 1999) are likely to operate in the sinus hair follicle as well. Whether the presence of glutamate signalling molecules associated with Merkel cell-neurite complexes is a part of that system remains to be determined.

We have shown that MK-801, a blocker of the ionotropic NMDA channel blocks the responsiveness of the St I receptors in isolated vibrissae (Senok et al. 2001), but we have since been unable to show a clear effect of other classical NMDA and non-NMDA antagonists on the function of the receptor. This suggests that either the pharmacological profile of the NMDA receptor in these mechanoreceptors is different from that in the CNS, or that the MK-801 is acting via a different mechanism.

In conclusion, the appropriate signalling molecules for glutamate transmission appear to be present in the Merkel cell-nerve terminal complexes, but it remains to be shown whether glutamate (or another excitatory amino acid) is involved in Merkel cell mechanosensory function.

Acknowledgements

This work was supported by the Wellcome Trust (ref: 061023) and the University of Stirling.

References

Chan E, Yung WH, Baumann KI (1996) Cytoplasmic Ca^{2+} concentrations in intact Merkel cells of an isolated, functioning rat sinus hair preparation. Exp Brain Res 108:357–366

Fagan BM, Cahusac PM (2001) Evidence for glutamate receptor mediated transmission at mechanoreceptors in the skin. Neuroreport 12:341–347

Genever PG, Maxfield J, Kennovin D, Maltman J, Bowgen CJ, Raxworthy MJ, Skerry TM (1999) Evidence for a novel glutamate-mediated signalling pathway in keratinocytes. J Inv Dermatol 112:337–342

Ogawa H (1996) The Merkel cell as a possible mechanoreceptor cell. Progr Neurobiol 49:317–334

Senok SS, Baumann KI (1997) Functional evidence for calcium-induced calcium release in isolated rat vibrissal Merkel cell mechanoreceptors. J Physiol 500:29–37

Senok SS, Baumann KI, Halata Z (1996) Selective phototoxic destruction of quinacrine-loaded Merkel cells is neither selective nor complete. Exp Brain Res 110:325–334

Senok SS, Baumann KI, Cahusac PM (2001) MK-801 depression of vibrissal slowly adapting mechanoreceptor responses. J Physiol 536P:38

Yamashita Y, Ogawa H (1991) Slowly adapting cutaneous mechanoreceptor afferent units associated with Merkel cells in frogs and effects of direct currents. Somatosens Mot Res 8:87–95

Transduction in Merkel Cell Mechanoreceptors – What Is the Job of the Merkel Cell?

Klaus I. Baumann[1] and Solomon S. Senok[2]

[1]Department of Functional Anatomy, University of Hamburg, UKE, Martinistr. 52, 20246 Hamburg, Germany

[2]Dept. of Physiology, College of Medicine and Medical Sciences, Arabian Gulf University, P.O. Box 22979, Manama, Bahrain

Summary

Most cutaneous mechanoreceptors are known to be primary mechanoreceptors. Mechano-electric transduction occurs in the terminal of the afferent nerve fibre, while the surrounding structures, e.g. lamellae in Meissner and Vater-Pacini corpuscles assist in directing the mechanical stimulus towards cytoplasmic protrusions extending from the nerve terminal. The latter are believed to be the site of mechanically gated ion channels. Merkel cell mechanoreceptors are different in the sense that special cells – Merkel cells – make close contact with nerve terminals showing dense core granules in the adjacent part of the Merkel cell cytoplasm and synapse-like membrane specializations in the contact area. Cytosplasmic protrusions are regularly seen extending from the Merkel cells rather than the nerve terminals. This has led to the hypothesis that the Merkel cell may be the mechanotransducer proper employing synaptic transmission towards the nerve terminal.

Various substances as possible candidates for the postulated synaptic transmission were found in the dense core granules of Merkel cells. However, until recently, there has been no conclusive evidence for neurotransmitter function of any of these substances between Merkel cell and nerve terminal and no direct electrophysiological measurements of receptor potentials have so far been successful, due to the location of these receptors between epidermis and dermis making them virtually inaccessible for microelectrode recordings.

Recently, mechanically induced increases in intracellular calcium of Merkel cells have provided evidence for mechanically gated calcium channels in Merkel cells and calcium-induced calcium release from intracellular stores within the Merkel cell. Glutamate receptor antagonists have been shown to reduce static responses of Merkel cell mechanoreceptors. Thus, it appears that slowly maintained adapting responses of Merkel cell mechanoreceptors involve a mechano-electric transduction process in Merkel cells and glutamatergic neurotransmission to the associated nerve terminals. However, the specific glutamate receptor still remains to be identified.

Introduction

Merkel's original description of "touch corpuscles" left no doubt that he considered these complexes as mechanoreceptors (Merkel 1875). There is ultrastructural evidence that Merkel cells have synaptic contacts with the associated nerve terminals from afferent neurons of cranio-spinal ganglia (Iggo and Muir 1969; Chen et al. 1973). These complexes are known to serve as mechanoreceptors (Iggo and Findlater 1984). The function of a mechanoreceptor is to convert mechanical stimuli into nerve action potentials. In all locations, Merkel cells are positioned, relative to the nerve terminal, on the side that normally receives mechanical stimulation (Halata and Baumann 2000). The surface of Merkel cells is equipped with protoplasmic protrusions anchoring them between keratinocytes (Iggo and Muir 1969) which has led to comparisons with inner ear hair cells (Iggo and Findlater 1984; Baumann et al. 1990), where bending of the hairs opens mechanically gated ion channels, resulting in depolarization, an increase in free intracellular calcium, and finally transmitter release (Crawford et al. 1991). However, there has been only one single report in the literature on exocytosis of what was believed to be neurotransmitter from a Merkel cell to the adjacent nerve terminal (Chen et al. 1973). This report reviews the currently available experimental evidence on the role of the Merkel cell in the mechano-electric transduction process.

Results and Discussion

There has been a long-standing controversy over whether the mechanoelectric-transduction process occurs in the Merkel cell, or whether the Merkel cell only directs the mechanical stimulus towards the nerve terminal. While all other types of mechanoreceptors in the skin lack specialized cells that could have direct involvement in the mechano-electric transduction process, only Merkel cell-nerve terminal complexes in the skin (SA I receptors) and in sinus hairs (St I receptors) have specialized cells, i.e. Merkel cells positioned on that side of the nerve terminal that physiologically receives mechanical stimuli (Halata and Baumann 2000). Unfortunately, the location of these receptors makes direct electrophysiological recordings with microelectrodes impossible. This has led to a number of experimental approaches to indirectly examine the role of Merkel cells.

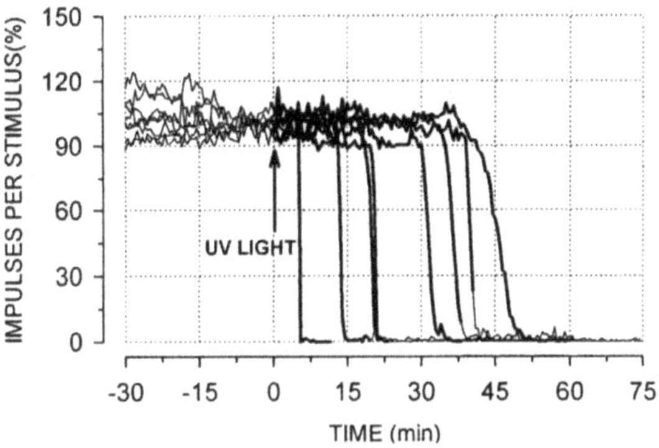

Fig. 1. Number of impulses from slowly adapting type I mechanoreceptors of isolated rat skin in response to standard mechanical stimuli (applied every 30 s) plotted against time. Responses to the last ten stimuli before the start of UV-irradiation were taken as reference (100%) and all responses are expressed as a percentage of this control value. Receptors were pretreated with quinacrine (10 μM for 30 min) shortly before the start of UV-irradiation at time 0. Receptor failure occurred on average after 27±4 min of UV-irradiation. Each *line* represents one of eight experiments. (Modified from Senok et al. 1996)

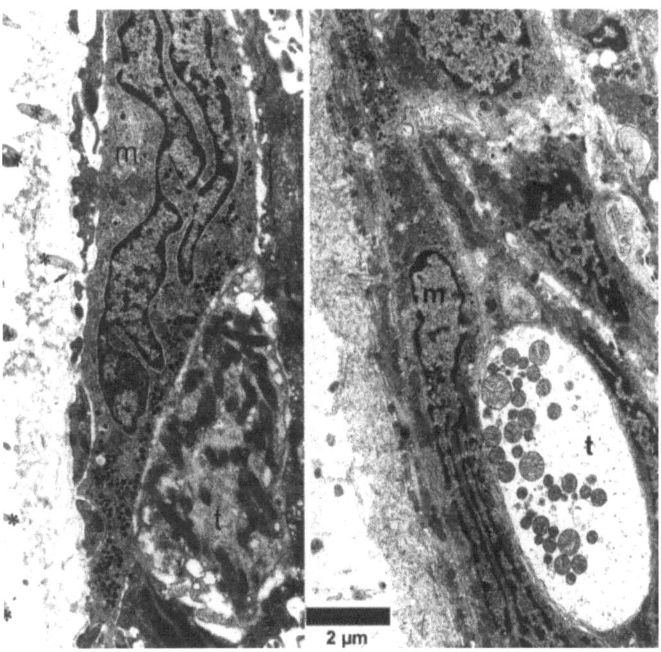

Fig. 2. Electron micrographs from Merkel cells (*m*) and nerve terminals (*t*) in rat sinus hairs. *Left* Control, *right* after quinacrine loading and UV-irradiation leading to receptor failure. (Modified from Senok et al. 1996)

Attempts to pharmacologically affect the responses of SA I receptors in touch domes were first carried out by Smith and Creech (1967), resulting in the conclusion "if a chemical neurotransmitter is present in ... it would seem to be different from presently known transmitters." The slowly adapting responses of these receptors, and characteristic impulse pattern of action potentials with rather irregular interspike intervals, was interpreted as evidence of a synaptic link (Iggo and Muir 1969; Horch et al. 1974). On the other hand, the high frequency of action potentials (up to 1500 Hz) with which these receptors could follow sinusoidal stimuli has been taken as an argument against synaptic transmission usually involving a synaptic delay of 0.5–1 ms (Gottschaldt and Vahle-Hinz 1981). Various peptides were found in the dense-cored granules of Merkel cells (Hartschuh et al. 1979; Hartschuh and Weihe 1988; Cheng Chew and Leung 1991; for a detailed review see English et al. 1992) and considered as potential neurotransmitters. However, naloxone had no effect on the responses of St I receptors to mechanical stimuli. Thus, met-enkephaline could not act as neurotransmitter between the Merkel cell and nerve terminal (Gottschaldt and Vahle-Hinz 1982).

In another experimental approach, Findlater et al. (1987) examined the effect of hypoxia on SA I responses. The authors showed that at the time when these receptors failed to respond to mechanical stimuli, the afferent nerve fibres were still conducting electrically elicited action potentials. Electron microscopic controls showed a significant reduction of dense-cored granules in Merkel cells at the time of receptor failure, suggesting that release of neurotransmitters from these granules is essential for normal receptor responses.

Diamond's group argued that Merkel cell mechanoreceptors still function after selective destruction of Merkel cells following quinacrine loading and bleaching with ultraviolet light (Diamond et al. 1988; Mills and Diamond 1995). There are contradictory reports concerning the effect of such procedures on the responses of Merkel cell mechanoreceptors (Ikeda et al. 1994; Senok et al. 1996). Figure 1 shows results from Senok et al. (1996): SA I receptors loaded with quinacrine and exposed to UV-light afterwards developed receptor failure within an average period of 27 min. At the time when receptors stopped responding to mechanical stimuli, electrical stimuli were still eliciting action potentials in the afferent nerve fibre. However, often the threshold for electrical stimuli increased rapidly within a few minutes after receptor failure and electron-microscopic examination of such UV-irradiated Merkel cell-nerve terminal complexes showed that the tissue damage was not selective, but extended to the nerve terminals as well, while leaving some Merkel cells relatively unharmed (Fig. 2). Thus, this experimental approach cannot answer the question of where the mechano-electric transduction process actually occurs (Senok et al. 1996).

It has long been established that an increase in free calcium is required for synaptic transmitter release (Katz and Miledi 1967). Thus, experiments have been carried out to study the effect of calcium channel blockers on receptor responses. Pacitti and Findlater (1988) observed dose-dependent reduction in responses of

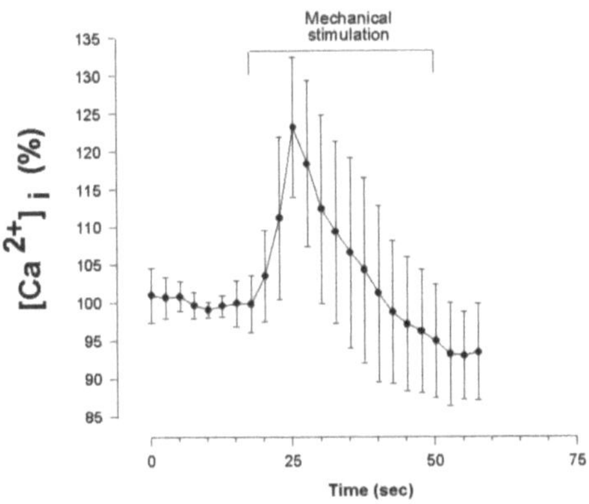

Fig. 3. Effect of direct mechanical stimulation via glass microprobes on free intracellular calcium concentration in Fluo-3 loaded Merkel cells of functionally intact isolated rat sinus hairs. Mean values ± SEM of seven experiments. (Modified from Chan et al. 1996)

SA I receptors after application of verapamil and these results were interpreted in support of the postulated synaptic transmission process. In patch clamp studies on isolated Merkel cells, Yamashita et al. (1992) demonstrated the presence of voltage-gated calcium channels similar to those found in other excitable cells, but could not find any mechanically gated ion channels.

Recently, dyes have become available that are taken up by living cells and emit fluorescent light, as such demonstrating the free intracellular calcium level (Grynkiewicz et al. 1985). These dyes allow examination of the intracellular calcium concentration in Merkel cells during mechanical stimulation, not only in single isolated cells, but also isolated receptors leaving the anatomical connections to the adjacent keratinocytes and nerve terminals intact (Baumann et al. 1996). Increases in intracellular calcium concentration in Merkel cells during mechanical stimulation had been observed in both rat St I receptors (Chan et al. 1996) and hamster SA I receptors (Tazaki and Suzuki 1998). Consistent increases in free intracellular calcium of Merkel cells can be observed during direct mechanical stimulation of Merkel cells with microprobes (Fig. 3), as well as during slight swelling of Merkel cells caused by exposure to hyposmotic solutions (Fig. 4). In spite of mechanical stimulation, no such increase is seen if the extracellular fluid is calcium-free (Fig. 4) or, if amiloride (known to block mechanosensitive ion channels; Hamill et al. 1992) is added to the solution. Thus, mechanical stimulation causes an influx of calcium into Merkel cells, which in turn appears to trigger further release of calcium from intracellular stores (Calcium-induced calcium release; Senok and Baumann 1997). Further studies by Tazaki and his group (Tazaki et al., this Vol.) suggest that L- and P/Q-type calcium channels are regularly found in Merkel cells, while N-type channels are found in the majority, but not all Merkel cells.

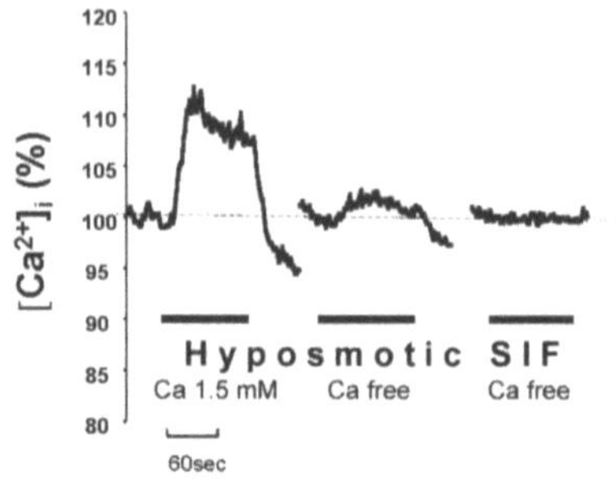

Fig. 4. Effect of hyposmotic swelling on free intracellular calcium concentration in calcium (1.5 mM) and calcium-free solution. Mean values of eight experiments

In the long search for neurotransmitter candidates at the postulated synapse between Merkel cell and nerve terminal, recent experimental evidence shows that the broad spectrum ionotropic glutamate receptor antagonist kynurenate severely depresses responses of St I receptors especially during the static phase (Fagan and Cahusac 2001) while dynamic responses were much less affected. This finding is in support of the two-receptor-sites hypothesis (Ogawa 1996) suggesting that the nerve terminal is able to produce a transient, rapidly adapting discharge during the dynamic phase of stimulation, while the typical slowly adapting impulse pattern of these receptors in response to maintained mechanical stimuli relies on a transduction mechanism residing in the Merkel cell and glutamatergic transmission to the nerve terminal.

Positive immunoreaction for subunits of NMDA receptors in Merkel cell-nerve terminal complexes were found by Senok et al. (this Vol.) in line with the assumed glutamatergic transmission. However, their light microscope studies could not unambiguously identify whether these NMDA receptors were located on the membrane of the nerve terminal or the Merkel cell. Tachibana and Nawa (this Vol.) demonstrated beta-type phospholipase C (PLCβ) in nerve terminals associated with Merkel cells known to be linked to metabotropic glutamate receptors (Pin and Duvoisin 1995). While this finding generally supports the assumption of glutamatergic transmission in Merkel cell mechanoreceptors, it leaves the question wide open which type of glutamate receptor is involved in this location. Further experiments with different glutamate receptor blockers by Cahusac (this Vol.) did not result in clear identification of a particular type of glutamate receptor. Thus, in spite of significant progress in recent years, a clear picture of the specific role of Merkel cells within the mechano-electric transduction process is still lacking.

Acknowledgements

Part of this work was carried out in the Department of Physiology at the Chinese University of Hong Kong with support by the Research Grants Council of Hong Kong.

References

Baumann KI, Hamann W, Leung MS (1990) Acute effects of neomycin on slowly adapting type I and type II cutaneous mechanoreceptors in the anaesthetized cat and rat. J Physiol 425:527–544

Baumann KI, Chan E, Halata Z, Senok SS, Yung WH (1996) An isolated rat vibrissal preparation with stable responses of slowly adapting mechanoreceptors. Neurosci Lett 213:1–4

Chan E, Yung WH, Baumann KI (1996) Cytoplasmic Ca^{2+} concentrations in intact Merkel cells of an isolated, functioning rat sinus hair preparation. Exp Brain Res 108:357–366

Chen SY, Gerson S, Meyer J, Holzer FJ (1973) The fusion of Merkel cell granules with a synapse-like structure. J Invest Dermatol 61:290–292

Cheng Chew SB, Leung PY (1991) Immunocytochemical evidence of a met-enkephalin-like substance in the dense-core granules of mouse Merkel cells. Cell Tissue Res 265:611–614

Crawford AC, Evans MG, Fettiplace R (1991) The actions of calcium on the mechano-electrical transducer current of turtle hair cells. J Physiol 434:369–398

Diamond J, Mills LR, Mearow KM (1988) Evidence that the Merkel cell is not the transducer in the mechanosensory Merkel cell – neurite complex. Progr Brain Res 74:51–56

English KB, Wang ZZ, Stayner N, Stensaas LJ, Martin H, Tuckett RP (1992) Serotonin-like immunoreactivity in Merkel cells and their afferent neurons in touch domes from the hairy skin of rats. Anat Rec 232:112–120

Fagan BM, Cahusac PMB (2001) Evidence for glutamate receptor-mediated transmission at mechanoreceptors in the skin. Neuroreport 12:341–347

Findlater GS, Cooksey EJ, Anand A, Paintal AS, Iggo A (1987) The effects of hypoxia on slowly adapting type I (SAI) cutaneous mechanoreceptors in the cat and rat. Somatosens Res 5:1–17

Gottschaldt KM, Vahle-Hinz C (1981) Merkel cell receptors: structure and transducer function. Science 214:183–186

Gottschaldt KM, Vahle-Hinz C (1982) Evidence against transmitter function of met-enkephalin and chemosynaptic impulse generation in "Merkel cell" mechanoreceptors. Exp Brain Res 45:459–463

Grynkiewicz G, Poenie M, Tsien RY (1985) A new generation of Ca^{2+} indicators with greatly improved fluorescence properties. J Biol Chem 260:3440–3450

Halata Z, Baumann KI (2000) Topography of nerve terminals in Merkel nerve endings in mammals. In: Suzuki H, Ono T (eds) Merkel cells, Merkel cell carcinoma and neurobiology of the skin. Elsevier, Amsterdam, pp 33–42

Hamill OP, Lane JW, McBride DW (1992) Amiloride – a molecular probe for mechanosensitive channels. TiPS 13:373–376

Hartschuh W, Weihe E (1988) Multiple messenger candidates and marker substances in the mammalian Merkel cell – axon complex: a light and electron microscopic immunohistochemical study. Progr Brain Res 74:181–187

Hartschuh W, Weihe E, Buchler M, Helmstaedter V, Feurle GE, Forssmann WG (1979) Met enkephalin-like immunoreactivity in Merkel cells. Cell Tissue Res 201:343–348

Horch KW, Whitehorn D, Burgess PR (1974) Impulse generation in type I cutaneous mechanoreceptors. J Neurophysiol 37:267–281

Iggo A, Muir AR (1969) The structure and function of a slowly adapting touch corpuscle in hairy skin. J Physiol 200:763–796

Iggo A, Findlater GS (1984) A review of Merkel cell mechanisms. In: Hamann W, Iggo A (eds) Sensory receptor mechanisms. World Scientific, Singapore, pp 117–131

Ikeda I, Yamashita Y, Ono T, Ogawa H (1994) Selective phototoxic destruction of rat Merkel cells abolishes responses of slowly adapting type I mechanoreceptor units. J Physiol 479:247–256

Katz B, Miledi R (1967) The timing of calcium action during neuromuscular transmission. J Physiol 189:535–544

Merkel F (1875) Tastzellen and Tastkoerperchen bei den Hausthieren und beim Menschen. Arch Mikr Anat 11:636–652

Mills LR, Diamond J (1995) Merkel cells are not the mechanosensory transducers in the touch dome of the rat. J Neurocytol 24:117–134

Ogawa H (1996) The Merkel cell as a possible mechanoreceptor cell. Prog Neurobiol 49:317–334

Pacitti EG, Findlater GS (1988) Calcium channel blockers and Merkel cells. Prog Brain Res 74:37–42

Pin JP, Duvoisin R (1995) The metabotropic glutamate receptors: structure and functions. Neuropharmacology 34:1–26

Senok SS, Baumann KI (1997) Functional evidence for calcium-induced calcium release in isolated rat vibrissal Merkel cell mechanoreceptors. J Physiol 500:29–37

Senok SS, Baumann KI, Halata Z (1996) Selective phototoxic destruction of quinacrine-loaded Merkel cells is neither selective nor complete. Exp Brain Res 110:325–334

Smith KR, Creech BJ (1967) Effect of pharmacological agents on the physiological responses of hair discs. Exp Neurol 19:477–482

Tazaki M, Suzuki T (1998) Calcium inflow of hamster Merkel cells in response to hyposmotic stimulation indicate a stretch-activated ion channel. Neurosci Lett 243:69–72

Yamashita Y, Akaike N, Wakamori M, Ikeda I, Ogawa H (1992) Voltage-dependent currents in isolated single Merkel cells of rats. J Physiol 450:143–162

Merkel Cell Carcinoma

Merkel Cell Carcinoma – A Short Review

Ingrid Moll[1] and Roland Moll[2]

[1]Universitätsklinikum Hamburg-Eppendorf, Klinik für Dermatologie und Venerologie, Martinistraße 52, 20146 Hamburg, Germany
[2]Institut für Pathologie der Philips-Universität Marburg, Klinikum Lahnberge, Baldingerstraße, 35033 Marburg, Germany

Summary

Merkel cell carcinoma represents the neuroendocrine carcinoma of the skin which mostly develops in UV-exposed areas of elderly people.

Its clinical appearance is blue to red nodules. In H and E staining, monomorphic round cells resembling lymphoma cells are detectable. Ultrastructurally, these tumor cells reveal the typical neuroendocrine granules and fibrous bodies. Immunohistochemistry detecting the intermediate filaments became a milestone in the diagnosis of Merkel cell carcinomas. Thus, the coexpression of cytokeratins nos. 8, 18, 19, and especially no. 20, together with neurofilament proteins is, within the skin, unique for Merkel cell carcinoma. In addition, the decoration of paranuclear plaques of intermediate filaments are highly characteristic as well as neuroendocrine markers such as NSE, PGP 9.5 and N CAM. The histogenesis of Merkel cell carcinoma still remains speculative. Merkel cell carcinoma shares structural features with normal Merkel cells, such as neurosecretory granules and cytokeratin 20, but is different concerning the presence of fibrous bodies and neurofilament proteins. One hypothesis is that a pluripotent basal keratinocyte may give rise to Merkel cell carcinoma.

Introduction

Merkel cell carcinoma was first described as trabecular carcinoma and thought to represent an eccrine sweat gland-derived carcinoma by Toker (1972). A few years later, the same authors identified the typical neuroendocrine granules within the tumor cells giving rise to the name Merkel cell carcinoma (Tang and Toker 1978).

Merkel cell carcinomas develop in the elderly (mean age 65–70 years) equally in both sexes. Preferentially, these tumors arise in UV-exposed skin, mostly on the face, neck and limbs. The classical clinical appearance is that of a blue, violet-to-red, dome-shaped nodule, mostly with an intact epidermal surface (Fig. 1). Ulceration may occur, but is restricted to late stages and large tumors. Roughly about 25% of the Merkel cell carcinomas are associated with other neoplasms, mostly skin or head and neck cancers or hematologic neoplasms (Brenner et al. 2001).

Histology

The cut surface is that of a soft, "marrowy" tumor. The H and E staining shows the tumor tissue within reticular dermis extending to the subcutaneous tissue. The epidermis, papillary dermis and the adnexal structures are spared. The histological appearance may vary considerably, but typically the tumor appears cellular and monomorphic, with small to intermediate-sized cells, and often resembles a lymphoma (Fig. 2).

Ultrastructure

At the electron microscopic level, the neuroendocrine granules (diameter 100 nm) and the prominent paranuclear whorls of intermediate filaments ("fibrous bodies") are important criteria (for review see Moll et al. 1986). In addition, intermediate filaments may make up a more or less developed cytoskeleton. The tumor cells are coupled by inconspicuous small cell junctions which are not yet well characterized, and rare desmosomes.

Immunohistochemistry

In the diagnosis of Merkel cell carcinoma, immunohistochemistry using antibodies to intermediate filament proteins and neuroendocrine markers became milestones during the last years. Among the intermediate filaments, the keratin polypeptides of the simple epithelial type (or low-molecular weight cytokeratins) nos. 8, 18, 19 and 20 are the most prominent ones, whereas keratins of stratified epithelia as present in keratinocytes are absent (Merot et al. 1986; Moll et al. 1986). Among these, cytokeratin 20 is the most important one because of its specificity for Merkel cells and Merkel cell carcinomas within the skin. Moreover, cytokeratin 20 is absent in neuroendocrine carcinomas of other organs, e.g., the lung, and thus became extremely valuable in the differential diagnosis of Merkel cell carcinomas (Table 1). Interestingly, the keratins within the tumor cells are arranged in plaque-like structures representing the paranuclear whorls (see above), but may also make up a fine fibrillar cytoskeleton (Fig. 3a). Particularly the former arrangement is quite typical of Merkel cell carcinomas and valuable in the diagnosis of the tumor (Table 1).

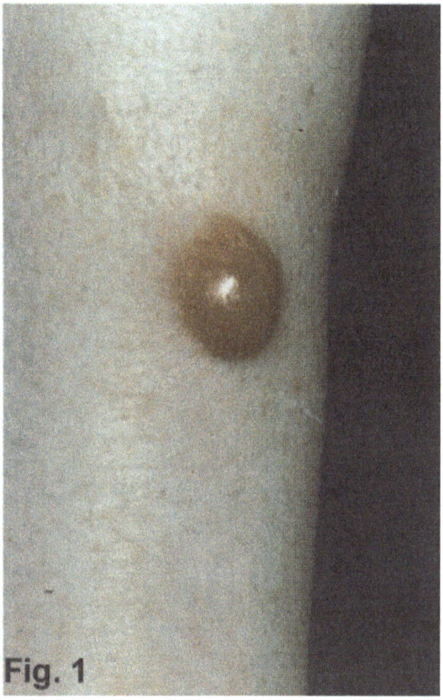

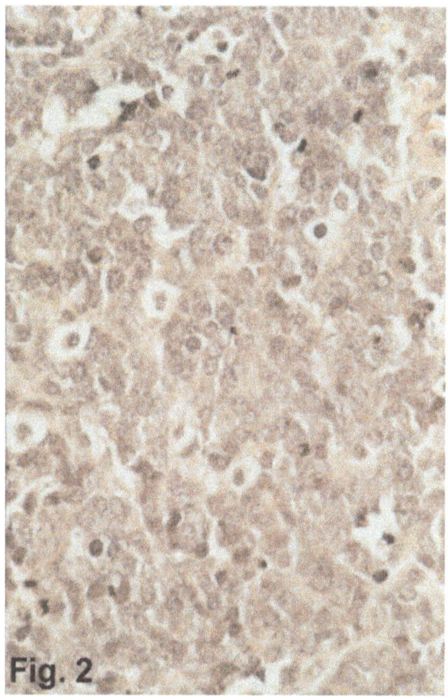

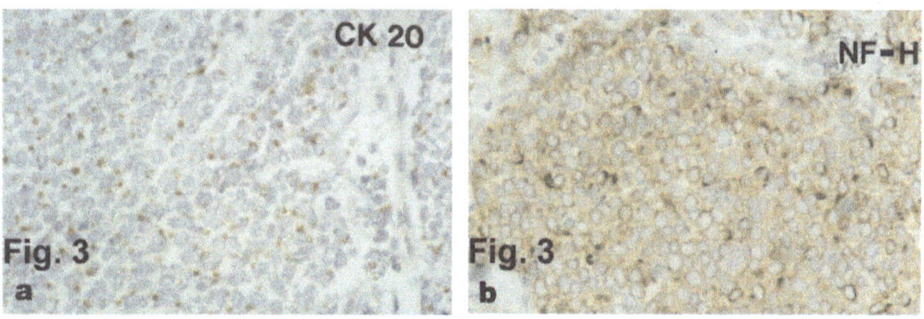

Fig. 1. A dome-shaped violet-colored Merkel cell carcinoma with intact skin surface

Fig. 2. H and E staining of a Merkel cell carcinoma showing uniform cells with large nuclei and scanty and poorly defined cytoplasm

Fig. 3. Immunoperoxidase staining of a Merkel cell carcinoma with antibodies to cytokeratin 20 (**a**) and neurofilament protein (large subunit NF-H; **b**). The paranuclear plaques are decorated by both antibodies, but cytokeratin 20 is detectable in nearly all tumor cells (**a**), whereas neurofilaments are only present in some (**b**). In addition, an irregular fibrillar cytoskeleton composed of cytokeratin 20 and neurofilaments is present

Neurofilaments are the second intermediate filament class present in most Merkel cell carcinomas (Fig. 3b), although often at lower amounts. The neurofilaments are mostly localized within the paranuclear whorls and to a lesser extent within the cytoskeleton. A further neuronal-type intermediate filament, peripherin, is often detectable in Merkel cell carcinomas (for review see Moll et al. 1997).

Table 1: Immunohistochemistry for diagnosis of Merkel cell carcinoma

	Immunoreactivity									
Tumor	Cytokeratin 8	Cytokeratin 20	Stratified epithelial Cytokeratins	Vimentin	Neurofilament proteins	NSE	Chromogranin A	S-100 protein	Leukocyte Common Antigen	Thyreoideal Transcription faktor (TTF-1)
Merkel cell carcinoma	+ (plaques)	+ (plaques)	-	-	+ (often plaques)	+	+/-	-	-	-
Malignant Lymphoma	-[1]	-	-	-	-	-/(+)	-	-	+	-
Malignant melanoma	-/+	-	-	+	-	-/+	-	+	-	-
Small cell carcinoma of the lung	+[2]	-	-	-[1]	-/+	+/-	-/+	-	-	+
Squamous cell carcinoma (various locations)	+/-	-	+	-/+	-/+	-	-	-	-	-

1 rarely positive cells

2 in some cases also plaques

The immunohistochemical detection of cytokeratin 20 in coexpression with neurofilaments, localized to paranuclear whorls, identifies the Merkel cell carcinoma nearly unequivocally.

Corresponding to the nature of these tumors, neuroendocrine markers are also expressed. These include neuron-specific enolase (NSE), chromogranin A (CgA), protein gene product 9.5 (PGP 9.5), synaptophysin and neuronal cell adhesion molecule (NCAM). Among these, NSE is consistently and broadly expressed in neuroendocrine tumors, but also occurs in nonneuroendocrine tumors of various organs and, therefore, is only useful in context with other markers. In contrast, the specificity of CgA for neuroendocrine differentiation is clearly superior. The negativity of Merkel cell carcinomas for thyroid transcription factor (TTF-1) is helpful for the exclusion of a small cell carcinoma of the lung (Kaufmann and Dietel 2000; Table 1). Various neuropeptides, such as VIP, calcitonin, ACTH, substance P, bombesin, gastrin, and somatostatin are detectable immunohistochemically in rare cases only and correlated clinical syndromes are hardly known.

An important etiological factor for Merkel cell carcinoma seems to be UV irradiation because of the typical UV mutations (C→T, CC→TT) often identified in the tumor cells (Popp et al. 2002). Immunosuppression is another important pathogenetic factor (Brenner et al. 2001).

The therapy of Merkel cell carcinoma includes wide excision, sentinel node biopsy, radiation, and in the case of metastases, various chemotherapy protocols (Tai et al. 2000; Lawenda et al. 2001). The rate of recurrences and lymph node metastases is high and the 5-year-survival is only about 30%.

Hypotheses on the histogenesis and the "cell of origin" of the Merkel cell carcinoma still remain speculative. The main argument supporting the Merkel cell as "cell of origin" is based on morphological and biochemical similarities, especially the common presence of neuroendocrine granules and cytokeratin 20. However, this may simply reflect differentiation pathways rather than histogenesis.

On the other hand, there are differences between Merkel cells and Merkel cell carcinomas. The most obvious are the fibrous whorls and the neurofilaments which are both absent from normal Merkel cells. Moreover, usually no connections between the Merkel cell carcinoma and the neighbouring epidermis and hair follicles are detectable. Another argument against the Merkel cell as "cell of origin" is the absence of mitoses in Merkel cells of humans (Vaigot et al. 1987; Moll et al. 1996). Analogous to the presumed embryonic development of human Merkel cells within the epidermis, which has been suggested by several groups (Compton et al. 1990; Moll et al. 1990; Kanitakis et al. 2000; Tachibana et al. 2000), it is conceivable that pluripotent basal keratinocytes ("stem cells") of the epidermis and adnexal epithelia may give rise to Merkel cell carcinoma, which is a true epithelial neoplasm. Pathogenetic factors, such as UV irradiation and immunosuppression may be important in tumor development (Brenner et al. 2001; Popp et al. 2002).

A variety of well characterized cell lines has been established from Merkel cell carcinomas (Leonard et al. 1993; Moll et al. 1994). In vitro, some tumor lines closely resemble the in vivo-tumor cells and thus will be useful to further clarify the nature of Merkel cell carcinomas. Moreover, these lines may be helpful in establishing new therapeutic approaches to improve the currently mostly fatal outcome for the patients.

References

Brenner B, Sulkes A, Radowsky E, Feinmesser M, Yukelson A, Bor-Haim, Katz A, Idelevitch E, Neumann A, Barhana M, Fenig E (2001) Second neoplasms in patients with Merkel cell carcinoma. Cancer 91:1358–1362

Compton C-C, Regauer S, Seiler G-R, Landry D-B (1990) Human Merkel cell regeneration in skin derived from cultured keratinocyte grafts. Lab Invest 63:233–241

Feinmesser M, Halpern M, Fenig E (1999) Expression of the apoptosis-related oncogenes bcl-2, bax, and p53 in Merkel cell carcinoma: can they predict treatment response and clinical outcome? Hum Pathol 30:1367–1372

Gould VE, Moll R, Moll I, Lee I, Franke WW (1985) Neuroendocrine (Merkel) cells of the skin: hyperplasias, dysplasias, and neoplasms. Lab Invest 52:334–353

Höfler H, Kerl H, Rauch H-J, Denk H (1984) New immunocytochemical observations with diagnostic significance in cutaneous neuroendocrine carcinoma. Am J Dermatopathol 6:525–530

Kanitakis J, Jullien D, De Boer B, Claudy A, Dubernard JM (2000) Regeneration of cutaneous innervation in a human hand allograft. Lancet 18:1738–1739

Kaufmann O, Dietel M (2000) Expression of thyroid transcription factor-1 in pulmonary and extrapulmonary small cell carcinomas and other neuroendocrine carcinomas of various primary sites. Histopathology 36:415–420

Lawenda BD, Thiringer JK, Foss RD, Johnstone PAS (2001) Merkel cell carcinoma arising in the head and neck. Am J Clin Oncol 24:35–42

Leonard JH, Bell JR, Kearsley JH (1993) Characterization of cell lines established from Merkel cell ("small-cell") carcinoma of the skin. Int J Cancer 55:803–810

Mérot Y, Margolis RJ, Dahl D, Saurat J-H, Mihm MC Jr (1986) Coexpression of neurofilament and keratin protein in cutaneous neuroendocrine carcinoma cells. J Invest Dermatol 86:74–77

Moll I, Lane AT, Franke WW, Moll R (1990) Intraepidermal formation of Merkel cells in xenografts of human fetal skin. J Invest Dermatol 94:359–364

Moll I, Bohnert E, Herbst C, Förster W, Moll R, Franke WW (1994) Establishment and characterization of two Merkel cell tumor cultures. J Invest Dermatol 102:346–353

Moll I, Kuhn C, Moll R (1995) Cytokeratin 20 is a general marker of cutaneous Merkel cells while certain neuronal proteins are present. J Invest Dermatol 104:910–915

Moll I, Zieger W, Schulz M (1996) Proliferate Merkel cells were not detected in human skin. Arch Dermatol Res 288:184–187

Moll R, Osborn M, Hartschuh W, Moll I, Mahrle G, Weber K (1986) Variability of expression and arrangement of cytokeratin and neurofilaments in cutaneous neuroendocrine carcinomas (Merkel cell tumors): immunocytochemical and biochemical analysis of twelve cases. Ultrastruct Pathol 10:473–495

Moll R, Moll I, Gould VE (1997) Neuroendocrine-Merkel cells of the skin and their neoplasms. In: Lechago J, Gould VE (eds) Bloodworth's endocrine pathology. Williams and Wilkins, Baltimore, pp 641–661

Popp S, Waltering S, Herbst C, Moll I, Boukamp P (2002) UV-B-type mutations and chromosomal imbalances indicate common pathways for the development of Merkel and skin squamous cell carcinomas. Int J Cancer 99:352–360

Tachibana T, Fujiwara N, Nawa T (2000) Postnatal differentiation of Merkel cells in the rat palatine mucosa, with special reference to the timing of peripheral nerve development and the potency of cell mitosis. Anat Embryol 202:359–367

Tai PTH, Yu E, Winquist E, Hammond A, Stitt L, Tonita J, Gilchrist J (2000) Chemotherapy in neuroendocrine/Merkel cell carcinoma of the skin: case series and review of 204 cases. J Clin Oncol 18:2493–2499

Tang CK, Toker C (1978) Trabecular carcinoma of the skin. Cancer 42:2311–2321

Toker C (1972) Trabecular carcinoma of the skin. Arch Dermatol 105:107–110

Vaigot P, Pisani A, Darmon YM, Ortonne JP (1987) The majority of epidermal Merkel cells are non-proliferative. A quantitative immunofluorescence analysis. Acta Derm Venerol 65:517–541

Histochemistry of Glycoconjugates in Merkel Cell Carcinomas

K Sames[1], U. Schumacher[1], R. Moll[3], I. Moll[4], Z. Halata[1], B. Asmus[1], E.J.M. van Damme[2] and W.J. Peumans[2]

[1]Institut für Anatomie, Universitätsklinikum Hamburg-Eppendorf, Martinistraße 52, 20246 Hamburg, Germany

[2]Laboratorium voor Fytopathologie en Plantenbescherming, Katholieke Universiteit Leuven, Willem de Croylaan 42, 3001 Leuven, Belgium

[3]Zentrum für Pathologie, Philipps-Universität Marburg, Baldingerstraße, 35033 Marburg, Germany

[4]Klinik und Poliklinik für Haut- und Geschlechtskrankheiten, Universitätsklinikum Hamburg-Eppendorf, Martinistraße 52, 20246 Hamburg, Germany

Summary

Glycoproteins and proteoglycans are involved in tumor genesis, progression, and metastatic activity.

Hence, glycoprotein and proteoglycan immunohistochemistry was applied to material of nine cases of Merkel cell carcinomas preserved in paraffin. Merkel cell tumors were identified by cytokeratin and neurofilament immunohistochemistry. Antibodies against laminin and type IV collagen were used to reveal basement membrane material.

An atypical distribution of laminin and type IV collagen with localization outside the basement membranes suggests high invasiveness. The reduced intensity of the cellular and cell membrane-associated reaction of heparan sulfate (HS) antibodies in comparison to normal epidermal cells may be a sign of low differentiation. The intensive reaction of antibodies against chondroitin (Ch)-type glycosaminoglycans with the tumor stroma and the pericellular substance in–and-around well-vascularized tumor septae may indicate high proliferation and migration activity of the cells. O-Linked glycans, typical for the mucins of most epithelia seem to be absent as shown by the lectin binding pattern of Merkel cell carcinoma cells, a fact underlining the undifferentiated state of the tumor cells. Mannose-specific lectins not applied before to Merkel cell tumors as well as HPA showed no reaction or only weak reactions with the tumor cells. The lectin reaction of the tumor stroma, was intermediate between the tumor cells and extra-tumoral stroma indicating an influence of tumor cells on the stromal constituents in the Merkel cell

carcinoma.
The findings point to a low degree of differentiation and high malignancy of the tumors. The methods used have proved to be of diagnostic and prognostic value and should be further standardized for such purposes.

Introduction

Glycoconjugate studies in normal Merkel cells by lectins have been performed earlier (Rosati et al. 1984). So-called Merkel cell tumors represent early metastasizing neoplasms of high grade malignancy (Sibley et al. 1985; Bergere et al. 1990; De-Cesare et al. 1990; Waibel et al. 1991). Its cells contain neuron-specific enolase, chromogranin, synaptophysin, various peptides in neuroendocrine granules, epithelial membrane antigen (E29), a cytoskeleton out of neurofilaments, and low molecular weight cytokeratins (Battifora and Silva 1986; Moll et al. 1986; Heenan et al. 1990). Lectins and proteoglycan antibodies have been applied to a number of different tumors including Merkel cell carcinoma (Zina et al. 1983; Pajor et al. 1986; Esko et al. 1988; Massone et al. 1988; Ramirez-Bosca et al. 1988; Olivero et al. 1990; Seiz et al. 1990; Jackson et al. 1991; Waibel et al. 1991; Iozzo and Cohen 1994; Penneys and Shapiro 1994; Timar et al. 1994; Schuhmacher 1995; Iida et al. 1996; Knutson et al. 1996; Burg et al. 1999; Engelberg 1999; Kalish et al. 1999; Kleeff et al. 1999; Stanley et al. 1999; Vlodvsky et al. 1999).

Methods

Material of Merkel cell tumors preserved in paraffin (five primary tumors, one local recurrent one, and three metastases) was provided by the Department of Pathology, University of Marburg and the Department of Dermatology, University of Hamburg. For the diagnosis of a Merkel cell carcinoma, monoclonal vimentin, cytokeratin (CK), and neurofilament (NF) antibodies were used.

Anti-type IV collagen or laminin antibodies served for the detection of basement membrane material. Specific monoclonal antibodies against HS, unsulfated chondroitin (C-0-S), chondroitin-4- and -6-sulfates (C-4-S, C-6-S) (Davidoff and Schulze 1990) were used for the detection of proteoglycans. Omission of the first antibodies or the chondroitinase treatment unmasking CS epitopes were used as negative controls, sections with articular cartilage and the mast cells of the tumor tissue as positive controls. Lectins used for histochemistry are shown in Table 1. Control reactions were performed in the presence of the inhibitory monosaccharides or following neuraminidase predigestion (Plendl et al. 1989).

For the type IV collagen, laminin and HS immunoreactions as well as the MAA reaction and PAS staining, only seven cases have been evaluated because the material available was limited.

Results

A positive PAS reaction is visible in the basement membranes, the connective tissue around the tumor nodes, the tumor septae and the tumor stroma. Near the surrounding tissue of tumor nodes or blood vessels, a PAS-positive pericellular network can be observed. Tumor cells are PAS-negative.

Type IV collagen is found in basement membranes as well as in the perivascular stroma and in some regions forms a network in the intercellular spaces. The laminin reaction is very similar.

In the cells and cell membrane-associated substances a reaction for HS is visible. The tumor stroma reacts less intensely. Staining of normal epidermis shows higher intensities in the cells and membrane-associated substance as compared to staining in tumors. With CS (C-4- and C-6-sulfated) antibodies, the reaction intensities are higher in the tumor connective tissue than in the extra-tumoral connective tissue. The most extended staining of the tumor stroma is observed using the C-6-S antibody. A network of pericellular substance containing C-4-S and C-6-S is present near to the connective tissue surrounding tumor nodules and near to the larger tumor septae containing blood vessels. It resembles the pericellular PAS, laminin or type IV collagen network (Table 1). There is no immunoreactivity with the C-O-S antibody. Controls without chondroitinase digestion also show a slight reaction, which may indicate the presence of incompletely glycosylated proteoglycans.

Most of the GalNac-specific lectins as well as the Fuc-specific UEA-I do not react with the tumor cells (Tables 2, 3). HPA also shows no reaction or only faint reactions with the tumor cells, even in material from metastases. In contrast to the CS antibodies, lectins react with higher intensities in the extra-tumoral connective tissue as compared to the tumor stroma and cells. All lectins show a reaction in the extra-tumoral tissue. The staining intensities of intra-tumoral connective tissue were between that of tumor cells and extra-tumoral connective tissue. The monosaccharide controls showed no staining at all or a drastic reduction of the staining intensities.

Tab. 1 Immunoreactivity of chondroitin type proteoglycan antibodies using tumor No 3 as an example

Structure	Chondroitin	Chondroitin-4-S	Chondroitin-6-S
Cytoplasm	-	+	-
Cell nucleus	-	+	-
Pericellular S.	-	+++	++++
Tumor Stroma	-	+	++++
Connect.T.l.	-	++	+++
Connect.T.d.	+	++	++
Adipose T.	+	+	+++

S. = substance, Connect. = connective, T. = tissue, l. = loose, d = dense
- = no reaction, + = uncertain reaction, ++ = well visible reaction, +++ = intense reaction, ++++ = very intensive reaction

Tab. 2 Lectins, their species of origin, their nominal sugar specificity and source used in this study

Abbreviation	Source	Nominal sugar	Supplier
HPA	Helix pomatia	GalNAc	Sigma
SBA	Glycine max	GalNAc	Sigma
IRA	Iris reticulata	GalNAc	In house
APA	Aegopodium podagraria	GalNAc	In house
UEA-I	Ulex europaeus	Fuc	Sigma
CMA	Chelidonium majus	GlucNAc	In house
WGA	Triticum aestivum	GlucNAc	Sigma
UDA	Urtica dioica	GlucNAc	In house
AUA	Allium ursinum	Man	In house
HHA	Hippeastrum hybrid	Man	In house
Con-A	Concanavalia ensiformis	Man/Glc	Sigma
LCA	Lens culinaris	Man/Glc	Sigma
ACA	Amaranthus caudatus	Gal	In house
MAA	Maackia amurensis	Sialic acid	Vector
SNA-I	Sambucus nigra	Sialic acid	In house

Tab. 3 Lectin binding to the tumor cells

Tumor Nr.	1	2	3	4	5	6	7	8	9
Lectin									
HPA	+	-	-	-	-	+	++	-	-
SBA	+	+	-	-	+	+	+	-	-
IRA	-	+	+	-	+	-	-	-	-
APA	-	-	-	-	-	-	+	-	+
UEA-I	-	-	-	-	-	-	-	-	-
CMA	++	+	++	+	+	++	++	+	-
WGA	+	++	+++	++	++	+++	++	++	+++
UDA	++	++	++	++	+++	+++	++	++	++
AUA	+	+	-	-	+	++	+	-	-
HHA	+++	+	++	+	+	++	+++	-	-
Con-A	++++	++++	+++	++++	++++	++++	++++	++++	++++
LCA	++	+++	++	+	+++	+++	+++	++	+
ACA	++	++	++	++	+	+++	+++	++	++
MAA	-	-	-	-	n.d.	-	-	n.d.	n.d
SNA-I	++	+	++	++	+	+++	++	+	+

Discussion

The deposition of laminin and type IV collagen outside basement membranes has been found to be an immunohistochemical marker of invasive capacity, especially in the case of laminin-5 (Felix et al. 1999; Kosmehl et al. 1999; Skyldberg et al. 1999; Stallmach et al. 1999; Ziober et al. 1999). GAG antibodies and the laminin antibody produce a similar distribution pattern which can be explained by the well-known binding of CS and HS to laminin (Blase et al. 1996; Knutson et al. 1996; Elias et al. 1999; Lugassy et al. 1999).

HS is reduced in a number of tumors showing a low degree of differentiation. Heparanase expression has been found to be a predispositon for malignancy (Timar et al. 1994; Knutson et al. 1996; Nackaerts et al. 1997; Burg et al. 1999). Our results indicate a reduced HS reaction in Merkel cell tumors. An upregulation or accumulation of Ch-type proteoglycans has been reported in the extracellular substances of several carcinomas (Jackson et al. 1991; Hinrichs et al. 1999; Ricciardelli et al. 1999; Stanley et al. 1999). The intensive immunoreactions with C-6-S and C-4-S antibodies in our study are in agreement with such studies (Murata 1980; Nigam et al. 1982; Romaris et al. 1994; Knutson et al. 1996). In general, the most pronounced proliferation activity of tumor cells takes place in regions with good vascularization and high CS-reactivity, indicating a possible role of CS in Merkel cell tumor proliferation.

Man-specific lectins (AUA, HHA), not tested before in the study of Merkel cell carcinomas, show no reaction or weak reactions with the tumor cells. Other new lectins (IRA, APA, ACA, SNA-I) confirm earlier reports with other lectins of the same specificity. There is only an occasional reactivity with GalNAc-specific and Fuc-specific lectins (Pajor et al. 1986; Waibel et al. 1991) pointing to the presence of O-linked mucus-type-glycoproteins typical for most of the differentiated epithelial cells (Strous and Dekker 1992). Thus, the glycosylation pattern of Merkel cell carcinoma cells differs substantially from the pattern of normally differentiated epithelial cells. The lack of reaction with HPA, known to react with tumor cells of epithelial origin (Thies et al. 2001), may be explained by an advanced dedifferentiation with loss of epitopes reacting with HPA.

The changed reaction intensities in the stroma and pericellular substance as compared to extra-tumoral connective tissue are in accordance with a control of the tumor stroma by neoplastic cells (Jackson et al. 1991).

References

Battifora H, Silva EG (1986) The use of antikeratin antibodies in the immunohistochemical distinction between neuroendocrine (Merkel cell) carcinoma of the skin, lymphoma, and oat cell carcinoma. Cancer 58:1040–1046

Bergere A, Lortholari A, Cornu P, Richard C, Bertrand G, Perrin A, Delpon A, Bernaille C (1990) Cutaneous neuroendocrine carcinoma. Apropos of a case. Rev Med Int 11:316–318

Blase L, Merling A, Engelmann S, Moller P, Schwartz-Albiez R (1996) Characterization of cell surface-expressed proteochondroitin sulfate of pre-beta Nalm-6 cells and its possible role in laminin adhesion. Leukemia 10:1000–1011

Burg MA, Pasqualini R, Arap W, Ruoslahti E, Stallcup WB (1999) NG2 proteoglycan-binding peptides target tumor neovasculature. Cancer Res 59:2869–2874

Davidoff M, Schulze W (1990) Combination of the peroxidase anti-peroxidase (PAP)- and avidin-biotin-peroxidase complex (ABC)-techniques: an amplification alternative in immunocytochemical staining. Histochemistry 93:531–536

De-Cesare A, Bononi M, Bianchi G, Atella F, Di-Stefano D, Mingazzini PL (1990) Merkel cell tumor. Ann Ital Chir 61:539–545

Elias MC, Veiga SS, Gremski W, Porcionatto MA, Nader HB, Brentani RR (1999) Presence of a laminin-binding chondroitin sulfate proteoglycan at the cell surface of a human melanoma cell Mel 85. Mol Cell Biochem 197:39–48

Engelberg H (1999) Actions of heparin that may affect the malignant process. Cancer 85:257–272

Esko JD, Rostand RS, Weinke JL (1988) Tumor formation dependent on proteoglycan biosynthesis. Science 241:1092–1096

Felix A, Rosa JC, Fonseca I, Cidado A, Soares J (1999) Laminin and collagen IV in pleomorphic adenoma and carcinoma ex-pleomorphic adenoma: an immunohistochemical study. Hum Pathol 30:964–969

Heenan PJ, Cole JM, Spagnolo DV (1990) Primary cutaneous neuroendocrine carcinoma (Merkel cell tumor) an adnexal epithelial neoplasm. Am J Dermatopathol 12:7–16

Hinrichs U, Rutteman GR, Nederbragt H (1999) Stromal accumulation of chondroitin sulphate in mammary tumours of dogs. Br J Cancer 80:1359–1365

Iida J, Meijne AM, Knutson JR, Furcht LT, Mc Carthey JB (1996) Cell surface chondroitin sulfate proteoglycans in tumor cell adhesion, motility and invasion. Semin Cancer Biol 7:155–162

Iozzo RV, Cohen I (1994) Altered proteoglycan gene expression and the tumor stroma. In: Jollès P (ed) Proteoglycans. Birkhäuser, Basel, pp 199–214

Jackson RL, Busch SJ, Cardin AD (1991) Glycosaminoglycans: molecular properties, protein interactions, and role in physiological processes. Physiol Rev 71:481–539

Kalish ED, Iida N, Moffat FL, Bourguignon LY (1999) A new CD44V3-containing isoform is involved in tumor cell growth and migration during human breast carcinoma progression. Front Biosci 4:A1–8

Kleeff J, Wildi S, Kumbasar A, Friess H, Lander AD, Korc M (1999) Stable transfection of a glypican-1 antisense construct decreases tumorigenicity in PANC-1 pancreatic carcinoma cells. Pancreas 19:281–288

Knutson JR, Iida J, Fields GB, McCarthy JB (1996) CD44/chondroitin sulfate proteoglycan and alpha2beta1 integrin mediate human melanoma cell migration on type IV collagen and invasion of basement membranes. Mol Biol Cell 7:383–396

Kosmehl H, Berndt A, Strassburger S, Borsi L, Rouselle P, Mandel U, Hyckel P, Zardi L, Katenkamp D (1999) Distribution of laminin and fibronectin isoforms in oral mucosa and oral squamous cell carcinoma. Br J Cancer 81:1071–1079

Lugassy C, Shahsafaei A, Bonitz P, Busam KJ, Barnhill RL (1999) Tumor microvessels express the beta-2 chain of laminin. Implication for melanoma metastasis. J Cutan Pathol 26:222–226

Massone L, Piccini R, Borgiani L, Borghi S, Pestarino A, Gambini C (1988) Neuroendocrine carcinoma of the skin (Merkel cell tumor). A clinical case with an ultrastructural and immunohistochemical study. G Ital Dermatol Venereol 123:219–223

Moll R, Osborn M, Hartschuh W, Moll I, Mahrle G, Weber K (1986) Variability of expression and arrangement of cytokeratin and neurofilaments in cutaneous neuroendocrine carcinomas (Merkel cell tumors): immunohistochemical and biochemical analysis of twelve cases. Ultrastruct Pathol 10:473–495

Murata K (1980) Structural differences of acidic glycosaminoglycans of leukocytes in leukemic states. Biochem Med 23:324–335

Nackaerts K, Verbeken E, Deneffe G, Vanderschueren B, Demedts M, David G (1997) Heparan sulfate proteoglycan expression in human lung-cancer cells. Int J Cancer 74:335–345

Nigam VN, Brailovsky CA, Bonaventure J (1982) Glycosaminoglycans and proteoglycans in neoplastic tissues and their role in tumor growth and metastasis. In: Varma RS, Varma R (eds) Glycosaminoglycans and proteoglycans in physiological and pathological processes of body systems. Karger, Basel, pp 354–371

Olivero G, Franchello A, Pacchioni D, Enrichens F, Mao P, Benedetto G (1990) A rare case of Merkel's tumor with intestinal metastases. Ann Ital Chir 61:277–280

Pajor L, Balazs M, Balogh J, Brittig F, Joos L, Linse R, Scholz M, Suba Z, Toth P (1986) Ultrastructural lectin histochemical and immunohistological observations on Merkel cell tumors. Path Res Pract 181:45–49

Penneys NS, Shapiro S (1994) CD44 expression in Merkel cell carcinoma may correlate with risk of metastasis. J Cutan Pathol 21:22–26

Plendl J, Schönleber B, Schmahl W, Schumacher U (1989) Comparison of the unmasking of lectin receptors by neuraminidase and by enzyme-free buffer alone. J Histochem Cytochem 37:1743–1744

Ramirez-Bosca A, Valcuende-Cavero F, Torres-Peris V, Castello-Rodellas A (1988) Lectin staining of a Merkel cell tumor. Med Cutan Ibero Lat Am 16:125–127

Ricciardelli C, Quinn DI, Raymond WA, McCaul K, Sutherland PD, Stricker PD, Grygiel JJ, Sutherland RL, Marshall VR, Tilley WD, Horsfall DJ (1999) Elevated levels of peritumoral chondroitin sulfate are predictive of poor prognosis in patients treated by radical prostatectomy for early-stage prostate cancer. Cancer Res 59:2324–2328

Romaris M, Villena J, Molist A, Heredia A, Bassols A (1994) Ras transformation alters the composition of extracellular matrix proteoglycans in rat fibroblasts. Biochem Biophys Res Commun 200:925–932

Rosati D, Nurse CA, Diamond J (1984) Lectin binding properties of the Merkel cell and other root sheath cells in perinatal rat vibrissae. Cell Tissue Res 236:373–381

Schumacher U (1995) Lectins and cancer – an old field revisited. In: Pusztai A, Bardocz S (eds) Lectins, biomedical perspectives. Taylor and Francis, Bristol, pp 211–224

Seiz V, Sames K, Schmiegelow P (1990) Histotopochemical quantification of glycosaminoglycans in melanoma and surrounding epidermis. Dermatol Mon Schr 176:745–755

Sibley RK, Dehner LP, Rosai J (1985) Primary neuroendocrine (Merkel cell) carcinoma of the skin. I. A clinicopathologic and ultrastructural study of 43 cases. Am J Surg Path 9:95–108

Skyldberg B, Salo S, Eriksson E, Aspenblad U, Moberger B, Tryggvason K, Auer G (1999) Laminin-5 as a marker of invasiveness in cervical lesions. J Natl Cancer Inst 91:1882–1887

Stallmach A, Orzechowski HD, Feldmann P, Riecken EO, Zeitz M, Herbst H (1999) 32/67-kD laminin receptor expression in human colonic neoplasia elevated transcript levels correlate with the degree of epithelial dysplasia. Am J Gastroenterol 94:3341–3347

Stanley MJ, Stanley MW, Sanderson RD, Zera R (1999) Syndecan-I expression is induced in the stroma of infiltrating breast carcinoma. Am J Clin Pathol 112:377–383

Strous GJ, Dekker J (1992) Mucin-type glycoproteins. Crit Rev Biochem Mol Biol 27:57–92

Thies A, Moll I, Berger J, Schumacher U (2001) Lectin binding to cutaneous malignant melanoma: HPA is associated with metastasis formation. Br J Cancer 84:819–823

Timar J, Diczhazy C, Bartha I, Ladanyi A, Tarcsafalvi A, Lapis K (1994) PMA induces shift from chondroitin to heparan sulphate on proteoglycans correlating with fibronectin adhesion of MDS human leukemia cells. Anticancer Res 14:1227–1231

Vlodavsky I, Friedman Y, Elkin M, Aingorn H, Atzmon R, Ishai-Michaeli R, Bitan M, Pappo O, Peretz T, Michal I, Spector L, Pecker I (1999) Mammalian heparanase: gene cloning, expression and function in tumor progression and metastasis. Nat Med 5:793–802

Waibel M, Richter K, Lengerken v K, Niedobitek F (1991) Merkel cell carcinoma (neuroendocrine carcinoma) in uncommon skin location. Immunohistochemical and lectin-histochemical findings. Zentralbl Pathol 137:140–150

Zina G, Zina AM, Bussolati G, Papotti M, Bundino S (1983) Merkel cell tumor (neuroendocrine carcinoma). Study of a case with ultrastructural and immunohistochemical findings. G Ital Dermatol Venereol 118:81–88

Ziober BL, Chen YQ, Ramos DM, Waleh N, Kramer RH (1999) Expression of the alpha7beta1 laminin receptor suppresses melanoma growth and metastatic potential. Cell Growth Differ 10:479–490

Gene Expression Profiling Reveals Two Distinct Subtypes of Merkel Cell Carcinoma

Mireille Van Gele[1], Glen Boyle[2], Anthony L. Cook[3], Tom Boonefaes[4], Pieter Rottiers[4], Nadine Van Roy[1], Anne De Paepe[1], Peter Parsons[2], J. Helen Leonard[3] and Frank Speleman[1]

[1]Center for Medical Genetics, Ghent University Hospital, Ghent, Belgium
[2]Queensland Institute of Medical Research, Herston, Brisbane, Queensland, Australia
[3]Queensland Radium Institute Research Unit, Herston, Brisbane, Queensland, Australia
[4]Department of Molecular Biomedical Research, Flanders Interuniversity Institute for Biotechnology and University of Ghent, Ghent, Belgium

Summary

Merkel cell carcinoma (MCC) is a rare aggressive neuroendocrine tumor of the skin. The genetic mechanisms underlying the development and tumor progression of MCC are poorly understood. We showed by comparative genomic hybridization analysis that the pattern of chromosomal abnormalities in MCC resembles that of small cell lung carcinoma (SCLC). Both tumors also share clinical and immunophenotypical characteristics. In addition, MCC cell lines can also be grouped, analogous to SCLC, into two different biological subgroups, namely Variant versus Classic MCC cell lines. In order to obtain more insight into the molecular pathogenesis of MCC and to find typical gene expression signatures associated with the phenotypically different subgroups of MCC cell lines, we determined the gene expression profiles of five Variant and five Classic MCC cell lines by the use of Atlas cDNA expression arrays. Supervised analysis allowed us to identify a set of 89 highly significant differentially expressed genes, which allowed classification of the MCC cell lines into the Variant and Classic subgroups. Genes mainly involved in cell cycle progression and cell proliferation showed higher expression levels in Variant MCC cell lines, mainly reflecting their more clinical aggressive behavior. Genes involved in signal transduction, neurotransmission and neuronal development showed a higher expression level in Classic MCC types associated with their more neuroendocrine and differentiated character. We assume that the differential expression levels of some of these genes reflect, analogous to SCLC, the different biological and clinical properties of Variant and Classic MCC phenotypes. Some of these genes could serve as useful prognostic markers and potential targets for the development of new therapeutic interventions specific for each subgroup.

Introduction

Merkel cell carcinoma (MCC) is a rare aggressive neuroendocrine skin tumor, mostly affecting elderly people. The disease occurs predominantly on the sun-exposed areas of the skin suggesting UV-exposure in its etiology (reviewed in Goessling et al. 2002). The number of MCC tumors and cell lines which have been cytogenetically investigated is relatively small (Van Gele et al. 2002). Although no recurrent translocations have been reported, cytogenetic analysis showed near-diploid karyotypes with complex structural changes in most cases, often presented as multi-way translocations. The distal part of the short arm of chromosome 1 is frequently affected and has been reported as the sole structural chromosome change in three MCC tumors (Gibas et al. 1994; Van Gele et al. 1998a; Van Gele et al. unpubl. results). In contrast to the rather inhomogeneous pattern of chromosomal changes described in cytogenetic studies, CGH analysis revealed a number of typical chromosomal imbalances including loss of chromosomal regions 3p, 5q, 8p, 10, 11q, 13q and 17p and gain of chromosomal regions 1, 3q, 5p and 8q (Van Gele et al. 1998b). Interestingly, the pattern of gains and losses resembled the pattern observed in previous CGH reports of small cell lung carcinoma (SCLC) (Ried et al. 1994; Levin et al. 1995; Petersen et al. 1997).

MCC and SCLC are both derived from neuroendocrine cells and share some clinical and immunohistochemical properties. In addition, similarities in morphology and growth characteristics were observed in cell lines derived from both MCC and SCLC. Comparable to SCLC cell lines, MCC cell lines can also be grouped into four different morphological classes (Type I to IV). In addition, different expression patterns of cytokeratins and neuroendocrine markers, such as chromogranin A and neuron-specific enolase, allowed further classification of these cell lines into Classic and Variant groups (Carney et al. 1985; Leonard and Bell 1997). Recent studies on the POU-IV family member *Brn-3c* and basic helix-loop-helix transcription factor *HATH1* have shown that lack of their expression was linked to loss of the neuroendocrine phenotype of Variant MCC cell lines (Leonard et al. 2002).

In contrast to MCC, numerous molecular genetic studies have been performed on SCLC which contributed to the understanding of SCLC pathogenesis (reviewed in Fong et al. 1999; Minna et al. 2002). In addition, histochemical markers and differentially expressed genes distinguishing Classic from Variant SCLC cell lines have been identified and led to an improved understanding of the underlying genetic basis responsible for the biological and clinical heterogeneity among small cell lung cancers (Broers et al. 1985; Zhang et al. 2000).

In order to obtain further insights into the molecular pathogenesis of MCC, the factors influencing cell line morphology, aggressiveness and neuroendocrine phenotype, we decided to determine the gene expression profiles of ten MCC cell lines using Atlas cDNA expression arrays containing 1891 unique genes involved in many cellular functions.

Material and Methods

Cell Lines
Merkel cell carcinoma cell lines MCC5, MCC6, MCC13, MCC14/1, MCC14/2 and MCC26 were established at the Queensland Radium Institute Laboratory, Queensland, Australia and have been described in detail previously by Leonard et al. (1993, 1995). MCC cell line UISO was described by Ronan et al. (1993), MKL-1 by Rosen et al. (1987). MKL-2 was established at the Robert H. Lurie Comprehensive Cancer Center, Illinois, USA and reported by Van Gele et al. (2002). T95–45 was established at the Center for Medical Genetics, Ghent, Belgium. Most cell lines were previously analyzed by CGH and/or M-FISH (Van Gele et al. 1998b, 2002). Table 1 provides an overview of the growth and morphological characteristics and classification of the MCC cell lines used in the differential gene expression analysis.

cDNA Array Hybridization
Total RNA from cell lines was extracted using the Atlas Pure Total RNA Labeling System (Clontech Laboratories, Palo Alto, CA, USA) and DNase I (Roche) treated. The quality and integrity of the DNase-treated RNA were checked by ethidium bromide agarose gel electrophoresis. Expression analysis was performed using the Atlas Human 1.2 (7850–1) and Atlas Human Cancer 1.2 (7851–1) nylon arrays (Clontech Laboratories, Palo Alto, CA, USA). Both filters contained 1176 of which 461 were present on both arrays. For each sample, 12.5 μg of total RNA was used in the cDNA probe synthesis with [α-^{32}P]dATP (NEN Life Science Products, Boston, MA, USA) and performed according to the Clontech Atlas cDNA Expression Arrays User Manual. Purification of the probe, hybridization and washes were done following the manufacturer's instructions (Clontech). Each cell line was simultaneously hybridized to both filters. After the washes, membranes were exposed for one to three nights to phosphoimager plates and scanned with a Storm Phosphoimaging System (Molecular Dynamics, Sunnyvale, CA, USA).

Analysis of cDNA Arrays
The scanned gel images were converted to 16bit TIFF-file format. Signal intensities were quantified using the VisualGrid software version 2.1 (http://www.GPC-Biotech.com). Analysis software was used for further primary data analysis. In short, the spot intensities were corrected for the local background signal intensity, followed by a spot quality control step to exclude spots influenced

by overshining effects of adjacent spots. Constitutive genes were selected (the 50% of spots showing the lowest coefficient of variation over all arrays) and used for normalization.

Expression Data Analysis

Genes with an expression value above background level in at least four of the analyzed samples were selected for further analysis. This resulted in a total of 1513 genes of which 410 were common genes.

We used the SAM algorithm (Significance Analysis of Microarrays, http://www.stat-class.stanford.edu/SAM/SAMServlet) which allows supervised identification of significantly expressed genes between predefined sample groups. Cluster and Treeview software were used for visualization of the data (Eisen et al. 1998).

Results

Identification of Differentially Expressed Genes in Variant Versus Classic MCC Cell Lines

In order to extract significant differentially expressed genes between Variant and Classic MCC cell lines, we adapted a supervised strategy. Briefly, two sample groups namely Variant (n=5) and Classic (n=5) MCC cell lines were predefined. Subsequently, a two-class SAM-analysis on the log transformed data matrix containing 1513 genes was performed. Thirty-three genes with an increased expression level in the Variant MCC cell lines and 56 genes with an elevated expression level in the Classic ones were identified. Hierarchical cluster analysis of these 89 genes clearly confirmed the classification of the MCC cell lines in their respective groups (data not shown). The genes with a higher expression in the Variant cell lines were particularly involved in cell cycle control and proliferation. Genes with kinase activities, but also genes encoding for ligand and voltage-gated ion channels, neuromediators, GDP/GTP exchangers and signal transduction receptors showed an increased expression level in the Classic MCC cell lines.

Discussion

Supervised analysis of array gene expression data revealed for the first time particular gene expression profiles for Variant and Classic MCC cell lines, respectively. We identified 89 highly significant differentially expressed genes of which 33 were higher expressed in the Variant MCC cell lines and 56 higher expressed in the Classic types. Upregulated genes in the Variant cell lines were involved in cell cycle control (*CCND1*, *CCNB1* and *G1 to S phase transition protein 1 homolog*) and cell proliferation (*HSP60*, *CSK2*, *MMP11*, *DBPA*, *MAPK9*, *FRA1* and *MAP kinase 38*). Some of these upregulated genes are possibly correlated with

Table 1. Characteristics of MCC cell lines used for gene expression profiling

MCC Cell line	Morphological type	Colony shape	Colony aggregation	Classification[a]
MCC5	I	3-d	Tight	Classic
MCC6	I	3-d	Tight	Classic
MCC13	IV	Flat	NA[b]	Variant
MCC14/1	IV	Flat	NA	Variant
MCC14/2	IV	Flat	NA	Variant
MCC26	IV	Flat	NA	Variant
UISO	IV	Flat	NA	Variant
MKL-1	III	2-d	Loose	Classic
MKL-2	III	2-d	Loose	Classic
T95-45	II	3-d	Loose	Classic

[a] Classic MCC cell lines have elevated expression levels of neuroendocrine markers including neurone-specific enolase and Chromogranin A and contain neurosecretory granules. Variant MCC cell lines have a selective loss of neuroendocrine markers including Chromogranin A and do not contain neurosecretory granules as evidenced by electron microscopy (Leonard and Bell 1997).

[b] Not applicable, adherent growing cell lines

the shorter doubling time and aggressive nature of the Variant MCC cell lines as illustrated by the high cloning efficiency in vitro of these cell lines and poor response to radiation (Leonard and Bell 1997). In addition, we observed high expression of *vimentin*, a mesenchymal marker, together with *FRA1* (*FOS-related antigen 1*). A tight correlation of *vimentin* and *FRA1* expression has also been recently found in highly invasive breast cancer cell lines pointing at a possible role in tumor progression and enhanced cell migration of these cancer cells (Zajchowski et al. 2001). These two genes could be significant prognostic markers for the more aggressive MCC Variant types. Increased expression of vimentin has also been previously observed by immunochemical studies in Variant SCLC cell lines (Broers et al. 1985, 1986) and as a result of a suppression subtractive hybridization experiment comparing a Variant to a Classic SCLC cell line (Zhang et al. 2000). These observations could point at a similar mechanism of tumor progression or metastatic properties between MCC and SCLC Variant phenotypes. A subset of genes with an increased expression level in the Classic cell lines are involved in signal transduction pathways leading to uncontrolled cell growth when overexpressed. This is exemplified by genes such as *MAPK3*, *MAPK7* and *MAPKK7* involved in the mitogen-activated protein (MAP) kinase pathway and genes such as *PI3-K p85 beta* and *PI4K-alpha* in the phosphatidylinositol 3-kinase (PI3 K) pathway. In addition, Classic cell lines showed increased expression of genes encoding for neuromediators and neurotransmitters and proteins involved in neuronal development such as lissencephalin and MARCKS-related protein

(*MLP*). This reflects the neuroendocrine and differentiated character of the Classic types. Ligand- and voltage-gated ion channels and receptors essential for neurotransmission were also upregulated. Some of these ion channels are known to play a role during mechanical stimulation of normal Merkel cell receptors (Baumann et al. 2000; Tazaki et al. 2000). Their specific function in MCC tumor cells has yet to be elucidated.

The differential expression levels of some of these markers reflect the different biological and clinical properties of Variant and Classic MCC phenotypes. Interestingly, SCLC tumors derived from morphological Variant cell lines are more aggressive and patients have a worse prognosis (Gazdar et al. 1985). Comparable to SCLC, lack or overexpression of Variant MCC markers could indicate a subset of more aggressive MCCs for which more intensive treatment and closer follow-up are warranted. Future investigations of overexpressed markers or deregulated pathways involved in Variant and Classic MCC cell lines could lead to potential targets for development of new therapeutic strategies specific for each subgroup.

In conclusion, we generated a gene expression-based classification of two biological and clinical distinct subgroups of MCC. This could result in a more selective therapeutic treatment and improvement of MCC patient outcome. Our study serves also as a first step to study differentially expressed genes involved in cell proliferation, signal transduction and neurotransmission in further detail, finally leading to more insight into the complex and heterogeneous biology of MCC.

References

Baumann KI, Senok SS, Chan E, Yung WH (2000) Calcium influx and calcium-induced calcium release in mechanically stimulated Merkel cells of rat sinus hair type I mechanoreceptors In: Suzuki H, Ono T (eds) Merkel cells, Merkel cell carcinoma and neurobiology of the skin. Proceedings of the 1st Symposium of the Japanese Society for Ultrastructural Curaneous Biology. Elsevier Science BV, Amsterdam, pp 73–81

Broers JL, Carney DN, de Ley L, Vooijs GP, Ramaekers FC (1985) Differential expression of intermediate filament proteins distinguishes classic from variant small-cell lung cancer cell lines. Proc Natl Acad Sci USA 82:4409–4413

Broers JL, Carney DN, Klein Rot M, Schaart G, Lane EB, Vooijs GP, Ramaekers FC (1986) Intermediate filament proteins in classic and variant types of small cell lung carcinoma cell lines: a biochemical and immunochemical analysis using a panel of monoclonal and polyclonal antibodies. J Cell Sci 83:37–60

Carney DN, Gazdar AF, Bepler G, Guccion JG, Marangos PJ, Moody TW, Zweig MH, Minna JD (1985) Establishment and identification of small cell lung cancer cell lines having classic and variant features. Cancer Res 45:2913–2923

Eisen MB, Spellman PT, Brown PO, Botstein D (1998) Cluster analysis and display of genome-wide expression patterns. Proc Natl Acad Sci USA 95:14863–14868

Fong KM, Sekido Y, Minna JD (1999) Molecular pathogenesis of lung cancer. J Thorac Cardiovasc Surg 118:1136–1152

Gazdar AF, Carney DN, Nau MM, Minna JD (1985) Characterization of variant subclasses of cell lines derived from small cell lung cancer having distinctive biochemical, morphological, and growth properties. Cancer Res 45:2924–2930

Gibas Z, Weil S, Chen ST, McCue PA (1994) Deletion of chromosome arm 1p in a Merkel cell carcinoma (MCC). Genes Chromosomes Cancer 9:216–220

Goessling W, McKee PH, Mayer RJ (2002) Merkel cell carcinoma. J Clin Oncol 20:588–598

Leonard JH, Bell JR (1997) Insights into the Merkel cell phenotype from Merkel cell carcinoma cell lines. Australas J Dermatol 38:91–98

Leonard JH, Leonard P, Kearsley JH (1993) Chromosomes 1, 11, and 13 are frequently involved in karyotypic abnormalities in metastatic Merkel cell carcinoma. Cancer Genet Cytogenet 67:65–70

Leonard JH, Dash P, Holland P, Kearsley JH, Bell JR (1995) Characterisation of four Merkel cell carcinoma adherent cell lines. Int J Cancer 60:100–107

Leonard JH, Cook AL, Van Gele M, Inglid KJ, Speleman F, Sturm RA (2002) Proneural and Proneuroendocrine transcription factor expression in cutaneous mechanoreceptor (Merkel) cells and Merkel cell carcinoma. Int J Cancer 101:103–110

Levin NA, Brzoska PM, Warnock ML, Gray JW, Christman MF (1995) Identification of novel regions of altered DNA copy number in small cell lung tumors. Genes Chromosomes Cancer 13:175–185

Minna JD, Roth JA, Gazdar AF (2002) Focus on lung cancer. Cancer Cell 1:49–52

Petersen I, Langreck H, Wolf G, Schwendel A, Psille R, Vogt P, Reichel MB, Ried T, Dietel M (1997) Small-cell lung cancer is characterized by a high incidence of deletions on chromosomes 3p, 4q, 5q, 10q, 13q and 17p. Br J Cancer 75:79–86

Ried T, Petersen I, Holtgreve-Grez H, Speicher MR, Schrock E, du Manoir S, Cremer T (1994) Mapping of multiple DNA gains and losses in primary small cell lung carcinomas by comparative genomic hybridization. Cancer Res 54:1801–1806

Ronan SG, Green AD, Shilkaitis A, Huang TS, Das Gupta TK (1993) Merkel cell carcinoma: in vitro and in vivo characteristics of a new cell line. J Am Acad Dermatol 29:715–722

Rosen ST, Gould VE, Salwen HR, Herst CV, Le Beau MM, Lee I, Bauer K, Marder RJ, Andersen R, Kies MS, Moll R, Franke WW, Radosevich JA (1987) Establishment and characterization of a neuroendocrine skin carcinoma cell line. Lab Invest 56:302–312

Tazaki M, Tazaki Y, Inoue T, Shimono M (2000) Calcium inflow of single Merkel cell in response to direct mechanical stimulation. In: Suzuki H, Ono T (eds) Merkel cells, Merkel cell carcinoma and neurobiology of the skin. Proceedings of the 1st Symposium of the Japanese Society for Ultrastructural Curaneous Biology. Elsevier Science BV, Amsterdam, pp 63–71

Van Gele M, Van Roy N, Ronan SG, Messiaen L, Vandesompele J, Geerts ML, Naeyaert JM, Blennow E, Bar-Am I, Das Gupta TK, van der Drift P, Versteeg R, Leonard JH, Speleman F (1998a) Molecular analysis of 1p36 breakpoints in two Merkel cell carcinomas. Genes Chromosomes Cancer 23:67–71

Van Gele M, Speleman F, Vandesompele J, Van Roy N, Leonard JH (1998b) Characteristic pattern of chromosomal gains and losses in Merkel cell carcinoma detected by comparative genomic hybridization. Cancer Res 58:1503–1508

Van Gele M, Leonard JH, Van Limbergen H, Van Roy N, Van Belle S, Cocquyt V, De Paepe A, Salwen H, Speleman F (2002) Combined karyotyping, CGH and M-FISH analysis allows detailed characterization of unidentified chromosomal rearrangements in Merkel cell carcinoma. Int J Cancer 101:137–145

Zajchowski DA, Bartholdi MF, Gong Y, Webster L, Liu HL, Munishkin A, Beauheim C, Harvey S, Ethier SP, Johnson PH (2001) Identification of gene expression profiles that predict the aggressive behavior of breast cancer cells. Cancer Res 61:5168–5178

Zhang L, Cilley RE, Chinoy MR (2000) Suppression subtractive hybridization to identify gene expressions in variant and classic small cell lung cancer cell lines. J Surg Res 93:108–119

Expression of Developmentally Regulated Transcription Factors in Merkel Cell Carcinoma

J.H. Leonard[1], A.L. Cook[1,2], M. Van Gele[3], F. Speleman[3] and R.A. Sturm[2]

[1]Queensland Radium Institute Research Unit, Queensland Institute of Medical Research, Brisbane, Queensland, Australia
[2]Institute for Molecular Bioscience, The University of Queensland, Brisbane, Queensland, Australia
[3]Center for Medical Genetics, Ghent University Hospital, Ghent, Belgium

Summary

We have examined a number of developmentally regulated transcription factors for expression in Merkel cells and MCC cell lines and demonstrated that their expression patterns may be prognostic in MCC. We have shown that human Merkel cells from adult scalp hair follicle and from neonatal foreskin epidermal sheets express Brn-3c and HATH1. In addition, results demonstrate that the novel Merkel binding activity complex MNF contains Brn-3c. Moreover, Classic lines which retain neuroendocrine phenotype, are slow growing in culture, grow in suspension and are thought to be less aggressive, retain Brn-2, Brn-3c and HATH1 expression, whereas Variant suspension lines which no longer express NE markers retain Brn-2 and Brn-3c expression, but lack HATH1. Further, Type IV Variant lines which grow as adherent monolayers have shorter doubling times, are more radiation-resistant and have higher cloning efficiencies in soft agar, all thought to be indicative of aggressive tumours, have reduced or no Brn-2 proteins and lack expression of Brn-3c and HATH1 transcription factors.

Skin Cancer in Australia

The settlement of tropical areas of the world with people of fair complexion has resulted in these regions having very high rates of skin cancer. This is the case in Australia (Armstrong and Kricker 2001); in particular, the State of Queensland (QLD) at latitude 27°S–11°S, is often called the Skin Cancer Capital of the World. Many of the first Caucasian settlers in QLD were of Celtic origin with fair skins which are much more sensitive to sunburn (Siskind et al. 2002; Youl et al. 2002). The Australian culture, which has evolved believing that to be tanned is to be healthy, has led to an increased exposure to the sun rather than avoidance. The reduction in working hours and changes in standards of dress has meant more leisure time spent at the beach or outdoors in much less clothing (Youl et al. 2002). It is hardly surprising, therefore, that the incidence of skin cancer has been steadily rising in Australia (Armstrong and Kricker 2001).

An increasing number of patients with Merkel cell carcinoma (MCC) have been referred to the Department of Oncology at the Royal Brisbane Hospital, formerly the Queensland Radium Institute (QRI) which is one of the main tertiary referral centres for radiation therapy in Queensland and it is this that initiated the research into MCC. Recent data from the Queensland Cancer Registry give the incidence and mortality of MCC and melanoma in Queensland for the period 1993–1999 (Table 1) and show that there is an upward trend in MCC incidence with a concomitant increase in the number of people dying from MCC.

Table 1. Incidence and mortality data for MCC and melanoma in Queensland

Year	Merkel cell carcinoma		Melanoma	
	Incidence	Mortality	Incidence	Mortality
1992	-	9	-	159
1993	15	6	1486	173
1994	14	3	1570	205
1995	30	9	1745	192
1996	26	5	1908	169
1997	37	12	2035	177
1998	50	10	2046	219
1999	35	15	1976	228

Etiology

The majority of MCC cases (41 of 65 cases) seen at QRI between 1981–1990 were from the head and neck region (Meeuwissen et al. 1995) with many of the patients having signs of long-term skin damage suggesting chronic sun exposure may be causal. In addition, there is now molecular evidence to support the view that sun exposure is involved in etiology, with recent reports demonstrating UV-B-type mutations in the *p53* and *(Ha)ras* genes in some MCC cell lines (Van Gele et al. 2000; Popp et al. 2002). Despite this, the process of neuroendocrine cell carcinogenesis, and basis of therapeutic responsiveness of these tumours remains poorly understood, limiting opportunities for improving clinical outcomes. In order to study this process, we established a number of MCC cell lines.

Merkel Cell Carcinoma Cell Lines

The series of MCC-derived cell lines were classified into two groups, those which retained their neuroendocrine phenotype and contained neurosecretory granules (Classic) and those in which no granules were observed by electron microscopy (Variant) following the classification for small cell lung cancer (SCLC; Carney et al. 1985; Gazdar et al. 1985). Within these groups, cell lines were further subdivided into four classes according to growth morphology: types I–III grew in suspension (Classic lines mainly confined to types I and II), whereas type IV grew as adherent monolayers, again following the classification for SCLC lines (Carney et al. 1985; Gazdar et al. 1985). Variant lines had shorter population doubling times in culture, were generally more radiation-resistant and tended to be more aneuploid (Leonard et al. 1993, 1995a,b). SCLC Variant lines form tumours more rapidly in nude mice, and patients with SCLC from whom Variant lines are established have a much poorer prognosis (Gazdar et al. 1985).

Developmental Regulatory Protein Expression

Merkel cells are thought to be a post-mitotic lineage, dividing Merkel cells having never been observed in normal skin (Moll et al. 1996). Therefore, cancer cells must be derived either from a stem cell which has not undergone differentiation, or must escape post-mitotic differentiation in order for the cancer to grow and divide. Data from skin grafts suggest that there is a stem cell since Merkel cells appear in grafts post-transplantation (Compton et al. 1990), but the cell has not yet been identified. In addition, many of the genetic changes seen in cancer cells cause alterations in genes involved in transcriptional regulation, for example, genes which are normally only switched on during embryogenesis (Monk and Holding 2001). One class of developmental regulators are the POU domain family of transcription factors, first recognised as a family by the identification of a bipartite DNA binding domain (Sturm and Herr 1988) made up of POU-specific and POU-homeo domains in the transcription factors Pit-1, Oct-1 and Unc86 which gave the family its name (Herr et al. 1988; Herr and Cleary 1995; for reviews see Rosenfeld 1991; Latchman 1999). The identification of Oct-1 was the first of what is now a large family of octamer binding proteins belonging to the POU class of transcription factors.

Octamer Transcription Factors

The octamer family of transcription factors was named according to the eight-base sequence of DNA (octamer) to which they bind (Sturm et al. 1988). Further members were identified in nuclear extracts prepared from mouse tissues (Schöler et al. 1989), with five cDNAs coding these proteins termed Brn-1–Brn-5 (for brain

1–5; He et al. 1989; Hara et al. 1992). In humans, complexes containing these proteins have been termed N-Oct proteins (for neuronal octamer binding proteins; Rosenfeld et al. 1991). Our work has concentrated on the Brn-3 proteins (POU class IV genes) which recognise a divergent octamer-related sequence and only bind the consensus octamer poorly, whereas the Brn-2 protein (POU class III gene) is able to bind both the consensus and the divergent target sites of the Brn-3 family (Latchman 1999).

Octamer Proteins and Cancer

Octamer protein expression in human tumours and tumour cell lines has been examined previously (Collum et al. 1992; Schreiber et al. 1992; Thomson et al. 1993, 1995), with astrocytoma and glioblastoma cell lines found to have a limited expression of octamer proteins. When compared to primary tumours which expressed most of the N-Oct proteins (N-Oct-2 through to N-Oct-5), the cell lines had a restricted pattern, with two astrocytomas lacking N-Oct-4 and all eight glioma lines expressing only very low levels of Oct-1 (Schrieber et al. 1994). Early work from our group showed that melanocytes, melanomas and melanoma cell lines expressed Oct-1, N-Oct-3 and N-Oct-5 (Thomson et al. 1993), however N-Oct-3 and N-Oct-5 are both products of the *brn-2* gene (Thomson et al. 1995; Atanasoski et al. 1997; Smith et al. 1998). In addition, when Brn-2 expression is ablated by the use of antisense RNA vectors in a tumorigenic human melanoma cell line MM96L, the cell line was no longer capable of forming tumours in nude mice (Thomson et al. 1995).

Expression of Octamer Proteins in MCC

Brn-2

Examination of a number of MCC cell lines, by electrophoretic mobility shift assay (EMSA), determined that there was less N-Oct-3 and N-Oct-5 in the MCC lines when compared with SCLC or melanoma cell lines, both of which have high levels of these proteins (Thomson et al. 1994). However, MCC suspension cell lines had more than MCC adherent lines which had little or no detectable N-Oct-3 or N-Oct-5 (Fig. 1A). We have further investigated this result by Western analysis using a polyclonal antibody to Brn-2 protein (Smith et al. 1998) which confirmed that MCC suspension lines had more Brn-2 than adherent lines (Fig. 1B). This result indicated that the Variant MCC suspension cell line, MCC1, had relatively less protein than MCC5 or MCC6 Classic lines when compared to GAPDH levels.

Brn-3a and Brn-3b

Both of the *brn-3a* and *brn-3b* genes have two protein isomers (Theil et al. 1993, 1995). The short form of Brn-3a, Brn-3a(s), occurs through alternative RNA

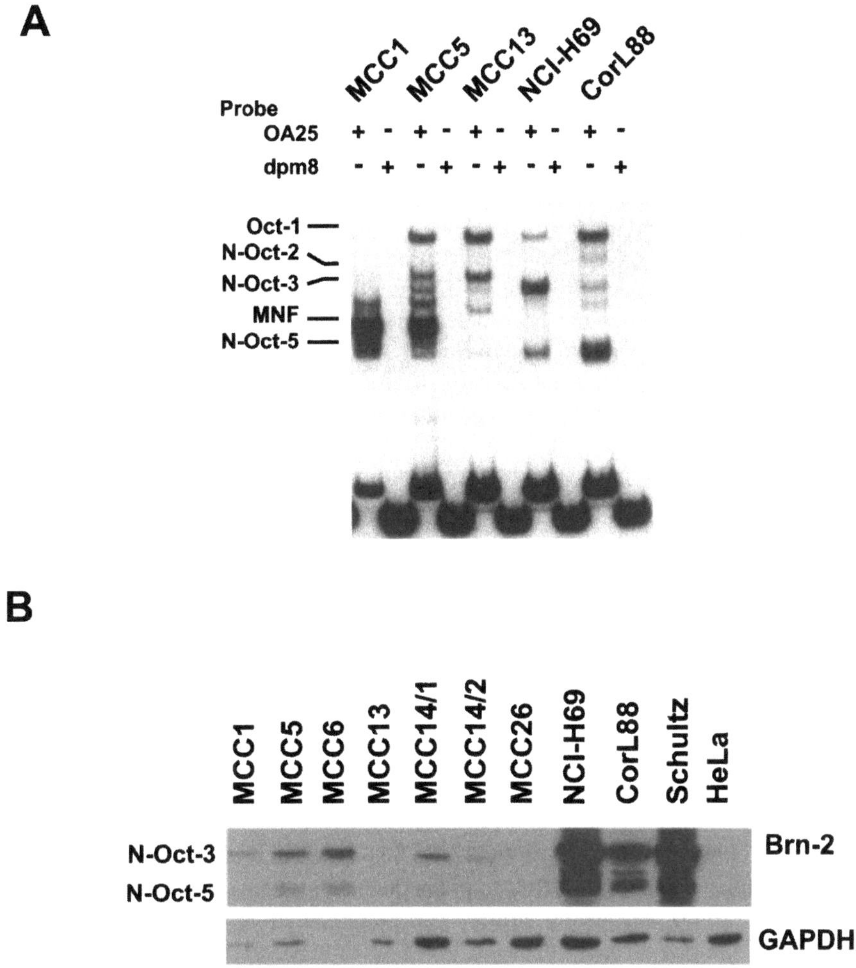

Fig. 1A, B. Expression of Brn-2 in MCC cell lines. **A** EMSA of MCC and SCLC cell lines with the non-consensus octamer probe (OA25) to which Brn-2 and Brn-3 proteins bind and as control a mutant octamer probe (dpm8) to which these proteins do not bind. Binding activities of Oct-1, the neuronal octamer protein N-Oct-2, the *brn-2* gene proteins N-Oct-3 and N-Oct-5 and MNF are indicated. **B** Western analysis for Brn-2 proteins of neuroendocrine cell lysates using a polyclonal antibody to Brn-2: MCC Variant suspension line MCC1, Classic suspension lines MCC5 and MCC6, Variant adherent lines MCC13, MCC14/1, MCC14/2, and MCC26, SCLC (NCI-H69, CorL88), Schultz a Ewing's sarcoma line and as control the cervical cancer line HeLa. Relative to SCLC and Ewing's sarcoma lines, all MCC lines had reduced amounts of N-Oct-3 and N-Oct-5 with Classic MCC lines having higher levels than Variant ones and only one adherent line MCC14/1 having detectable levels of protein splicing and lacks 80 N-terminal amino acids resulting in a truncated version of Brn-3a(l). Brn-3b(s) is derived from the use of an alternative translation initiation codon which produces a shorter protein, Brn-3b(s), but one which has an additional nine amino terminal residues not present in Brn-3b(l). Brn-3a has an important role in neural development and is able to induce expression of a number of synaptic vesicle proteins, including SNAP-25, essential for neurite outgrowth, and stimulate the expression of three neurofilament genes which provide support for neurite elongation and axon growth (reviewed by Latchman 1999). In the trigeminal ganglion, it is also required for correct neurotrophin receptor expression, critical for sensory neurone survival and differentiation (Huang et al. 1999). Its role in neuroendocrine cells has not been determined. Brn-3a is also able to induce expression of the anti-apoptotic protein Bcl-2 (Budhram-Mahadeo et al. 1999a). This is mediated by the N-terminal domain of Brn-3a(l), absent in Brn-3a(s) that is thought to interact with a neurone-specific co-activator protein. Interestingly, this domain is not required for Brn-3a to activate neurite outgrowth or neurofilament induction. The p53 tumour suppressor protein is able to inhibit *bcl-2* promoter activation by Brn-3a, possibly mediated by a p53 DNA binding site adjacent to the Brn-3a binding site in the *bcl-2* promoter (Budhram-Mahadeo et al. 1999a).

Brn-3b protein, in contrast to Brn-3a, inhibits neural outgrowth by active repression of genes induced by Brn-3a (Latchman 1999) and by forming inactive heterodimers with Brn-3a, it interferes with Brn-3a activation (Theil et al. 1995).

Mice nullizygous for *brn-3b*$^{-/-}$ have defective retinas, lacking approximately 70% of the retinal ganglion cells (RGCs), prompting Xiang (Xiang 1998) to suggest that these RGCs in the developing mouse retina require Brn-3b, while the remaining 30% can form independently of Brn-3b. However, recent work of Erkman and co-workers (Erkman et al. 2000), demonstrated that Brn-3b was at the head of a cascade required for axon pathfinding including genes required for cell – cell interactions and the actin-binding protein abLIM. In addition, nullizygous *brn-3b*$^{-/-}$ have a total loss of Brn-3a in the developing retina (Erkman et al. 1996; Trieu et al. 1999), suggesting Brn-3b also induces Brn-3a. Moreover, sequence analysis of the *brn-3a* gene revealed Brn-3 protein recognition sequences occur approximately 5 and 10 Kb upstream of the *brn-3a* transcription initiation site, thereby supporting a role for autoregulation of *brn-3a* by all or some Brn-3 family members (Trieu et al. 1999).

Brn-3a has been detected in neuroendocrine tumours, including SCLC, and rather than inducing oncogenesis, Brn-3a appears to be associated with the more aggressive phenotype of the tumours studied (Leblond-Francillard et al. 1997). Furthermore, *brn-3a* is over-expressed in human cervical cancer cell lines transformed by human papilloma virus oncogenes, and inhibition of *brn-3a* expression in these cells decreases their growth rate (Ndisdang et al. 1998). Northern blot analysis previously demonstrated *brn-3a* transcripts in six out of eight neuroepithelioma, four out of six Ewings sarcoma and one out of one SCLC cell lines, but not in eight neuroblastoma cell lines (Collum et al. 1992). Brn-3b has been implicated in sporadic, late-onset breast cancer and can repress BRCA-1 expression in breast cancer cells (Budhram-Mahadeo et al. 1999b).

We have been able to show expression of *brn-3a* and *brn-3b* in MCC cell lines by RT-PCR using specific primers first described by Xiang et al. (1995). We have used expression levels of Oct-1, the ubiquitously expressed POU domain family member, to establish relative amounts of cDNA from each cell line extract (data not shown). These were then utilised in three parallel PCR reactions with the specific primers for each of the Brn-3 family subtypes to give a semi-quantitative level of expression (Fig. 2). In a similar result seen for that with Brn-2, relative to SCLC lines, all MCC lines had lower levels of Brn-3a. Interestingly, the highest level was seen in MCC1, however the other Variant suspension cell line, MCC19, had only trace amounts as had all the other lines examined (Fig. 2, Panel A). Transcripts of *brn-3b* were seen in all MCC suspension lines with MCC1 and MCC19 Variant lines having higher relative levels then the Classic lines MCC5 and MCC6. Transcripts were not detected in the adherent lines or in any of the SCLC cell lines (Fig. 2B). Analysis of nuclear extracts of the different MCC cell lines by EMSA using the specific probe for Brn-3 family members did identify some complexes which were possibly due to binding of Brn-3a and Brn-3b.

Brn-3c

The Brn-3c transcription factor activates a similar array of genes to those activated by Brn-3a, but in addition, Brn-3c is involved in auditory system development (Erkman et al. 1996; Latchman 1999) as *brn-3c*$^{-/-}$ mice show no startled response to sharp sounds and have impaired balance, both indicating inner ear involvement. This has been shown to be due to a progressive loss of auditory and vestibular cells during development (Xiang et al. 1997, 1998). In humans, an 8-bp deletion in the homeodomain of Brn-3c, which results in low affinity binding to target sequences and impaired vestibular function, has implicated Brn-3c in a form of hereditary hearing loss (DFNA15; Vahava et al. 1998).

As shown in Fig. 2C, only MCC suspension cell lines contain *brn-3c* transcripts and we have also been able to demonstrate this by EMSA, where we have recently shown that a novel DNA binding protein identified in Merkel cell suspension cell lines, which we originally called Merkel Nuclear Factor (MNF; Fig. 1A; Thomson et al. 1994) is Brn-3c (Leonard et al. 2002). In addition, immunohistochemical analysis of tumour specimens and cell lines showed the cell lines reflected the reactivity seen in the biopsy material from which they were derived. Hence, as can be seen in Fig. 3, specimens from which adherent lines were generated had no nuclear reactivity to Brn-3c although there was some non-specific background staining in the cytoplasm (Leonard et al. 2002).

To determine whether this was a gain of expression of Brn-3c in MCC suspension cell lines, or a loss of expression in MCC adherent lines, we then examined Merkel cells in adult scalp hair follicles using confocal microscopy. Dual labelling techniques were employed to identify Merkel cells by reactivity to a monoclonal antibody raised against cytokeratin-20 (Dako), a specific marker for Merkel cells in the skin (Moll et al. 1995) and with a polyclonal antibody to Brn-3c (BAbCo). As shown in Fig. 3, Merkel cells at the base of the hair shaft had dual fluorescence (arrows) indicating that normal Merkel cells do indeed express Brn-3c. Brn-3c immunoreactivity has also recently been demonstrated in Merkel cells within the

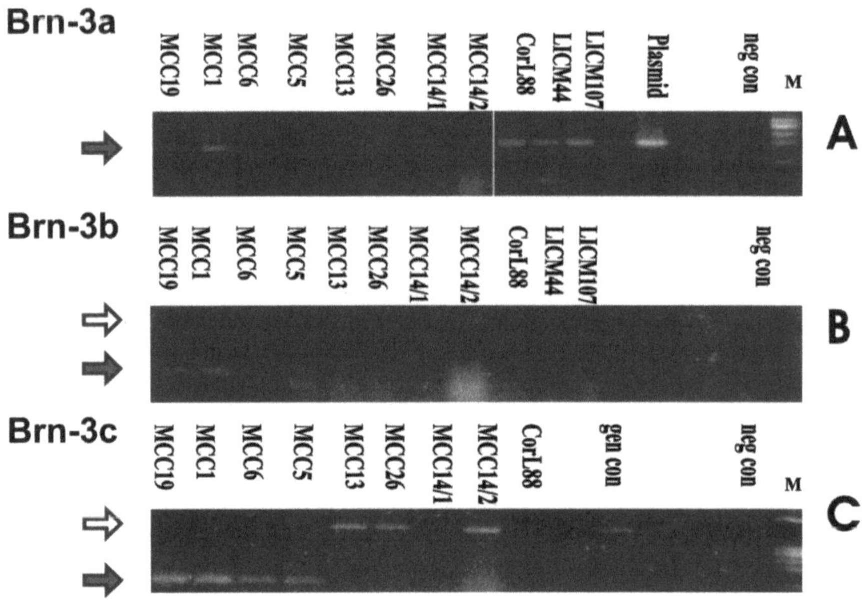

Fig. 2A–C. Expression of Brn-3 family genes in MCC. RT-PCR of cDNAs of MCC and SCLC lines with specific primers for Brn-3 genes. **A** Brn-3a primers showing low levels of expression in MCC lines relative to SCLC lines and no detectable levels in MCC adherent lines. Positive control is a plasmid vector containing the open reading frame of Brn-3a. **B** Brn-3b primers showing that only MCC suspension cell lines have detectable levels and **C** Brn-3c primers where again only MCC suspension lines have detectable levels. *Closed arrows* cDNA and *open arrows* genomic amplification. MCC suspension cell lines are MCC19, MCC1, MCC6 and MCC5. MCC adherent lines are MCC13, MCC26, MCC14/1 and MCC14/2. SCLC lines are CorL88, LICM44 and LICM107

basal layer of neonatal foreskin epidermis (Leonard et al. 2002) by immunofluorescence using a monoclonal antibody to chromogranin-A (CHM), a basic peptide found in neurosecretory granules and specific for Merkel cells within the epidermis (Hartschuh et al. 1989).

The Role of *HATH* Genes in Neural Development

Another family of genes which are transiently expressed in the chordotonal organs and photoreceptor cells, is the ATH transcription factors which are down-regulated as cells become post-mitotic. These genes encode a basic helix-loop-helix domain containing proteins which are conserved through evolution and most of the molecular mechanisms have been identified through work on the *Drosophila atonal* and the mouse *MATH1* ortholog. As yet, the human ortholog,

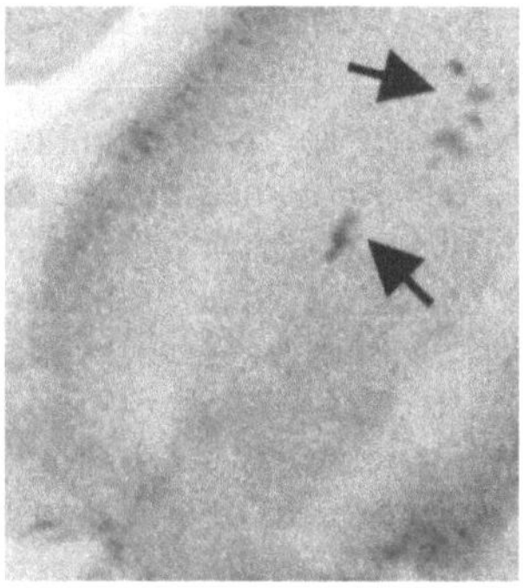

Fig. 3. Brn-3c in hair follicle Merkel cells. Dual label of scalp hair follicle by confocal microscopy. *Arrows* indicate cells at the base of hair shaft that have dual fluorescence for cytokeratin-20 (Dako), a specific marker for Merkel cells and Brn-3c, using Alexa conjugated secondary antibodies (molecular probes); ×200

HATH1, has not been extensively studied. Flies which lack *atonal* are blind and ataxic (Ben-Arie et al. 1996) and in mouse, ectopic expression of *MATH1* leads to neonatal death shortly after birth due to breathing difficulties with extensive effects on the brain and nervous system (Isaka et al. 1999). Expression is restricted to the external granular layer of the cerebellum, the dorsal neural tube, suggesting involvement in the onset of migration, and Merkel cells (see below). Recently, Bermingham and co-workers (Bermingham et al. 1999) showed HATH1 was required for hair cell development in the cochlea. Regulation of *MATH1* appeared to be through modular elements both 5′ and 3′ of the coding region and in mouse, including autoregulation by MATH1 (Helms et al. 2000). Other transcription factors known to be expressed in the embryonic mouse in the domains where MATH1 is seen are MSX and PAX families (Helms et al. 2000).

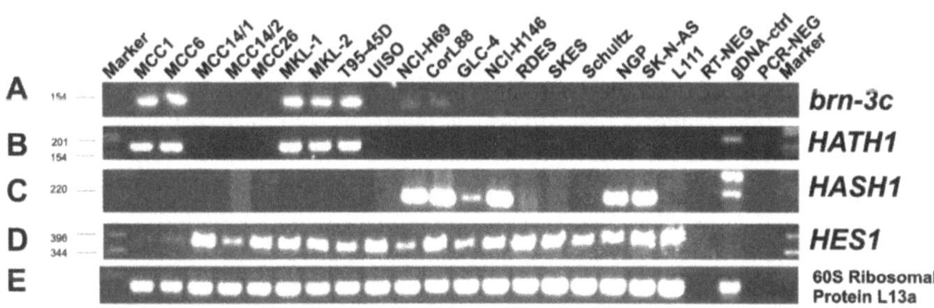

Fig. 4A-E. Expression of proneuroendocrine genes in NE carcinoma cell lines. RT-PCR with specific primers for **A** *brn-3c*,**B** *HATH1*, **C** *HASH1*, **D** *HES1* and as control 60S ribosomal RNA protein gene **E** *L13.* NE cell lines are: MCC – MCC1, MCC6, MCC14/1, MCC14/2, MCC26, MKL-1, MKL-2, T95-45D and UISO; SCLC – NCI-H69, CorL88, NCI-H146 and GLC-4: Ewing's sarcoma – RDES, SKES and Schultz; neuroblastoma – NGP and SK-N-AS and L111 is a large cell lung carcinoma cell line

Touch and Hearing Responses Are Related

In *Drosophila,* the cells which act as mechanoreceptors comprise the chordotonal organs found in the body wall, joints and antennae, providing the fly with balance and sensory information (Ben-Arie et al. 2000). Sound sensation is detected within the Johnston's Organ by ciliated sensory neurons (Eberl et al. 2000). Both sound and touch responses are eliminated in *atonal* mutant flies (Ben-Arie et al. 2000; Eberl et al. 2000). Recently, Ben-Arie and co-workers, using a mutant mouse containing *lacZ* in the coding region of the *MATH1* locus, were able to demonstrate that -gal positive cells in sections of hairy skin, footpad and vibrissae were Merkel cells. Examination of markers for Merkel cells in *MATH1* null mice revealed Merkel cells to be still present indicating MATH1 was required for their function, but not their genesis (Ben-Arie et al. 2000).

Examination of induced differentiation of the pluripotent human embryonal stem cell line NTERA-2 with retinoic acid caused a transient rise in *HATH1* expression which correlated with the expression of *nestin*, the first of a cascade of genes whose expression leads to cell differentiation and exit from the cell cycle (Przyborski et al. 2000).

To determine whether human Merkel cells contain HATH1, fluorescent microscopy was used to demonstrate reactivity in Merkel cells of neonatal foreskin (Leonard et al. 2002). A number of neuroendocrine-derived tumour lines were examined for expression of *HATH1* by RT-PCR, again using a semi-quantitative method to determine the relative amounts of transcripts against a control gene, the 60S ribosomal gene L13a. We found that *HATH1* transcripts were not present in any SCLC, Ewings' sarcoma or neuroblastoma cell line and were only present in a subset of MCC cell lines all of which grew as suspension cultures (Fig. 4B) and all of which expressed *brn-3c* (Fig. 4A). Immunohistochemical analysis of cell lines and tumour biopsies where available, demonstrated strong reactivity in some lines and their biopsies (Fig. 5). Of interest was the Variant MCC lines, as both suspension and adherent ones were negative for HATH1reactivity, even though the Variant suspension cell line, MCC1, had HATH1 transcripts and the biopsy from which it was derived had HATH1 positive cells (Figs. 4B, 6).

A recent report suggests that *MATH1* is not expressed in mouse lung (Yang et al. 2001), a finding consistent with our finding of no *HATH1* transcripts in SCLC cell lines (Fig. 4B). Rather, the bHLH transcription factor MASH1 ortholog of the *Drosophila* achaete-scute is required for pulmonary neuroendocrine cell development as in MASH1-deficient mice these cells fail to develop (Yang et al. 2001). In SCLC cells the human ortholog HASH1 has been shown to be linked with neuroendocrine phenotype (Ball et al. 1993). Interestingly, HASH1 expression can be repressed by expression of hairy-enhancer of split-1 (HES1) in SCLC cell lines with a concomitant and significant reduction of neuroendocrine markers in these cells (Chen et al. 1997). Examination of transgenic mice has shown that within the cochlea, mutation of HES1 leads to increased numbers of hair cells and that overexpression of HES1 blocks hair cell development, even in the presence of

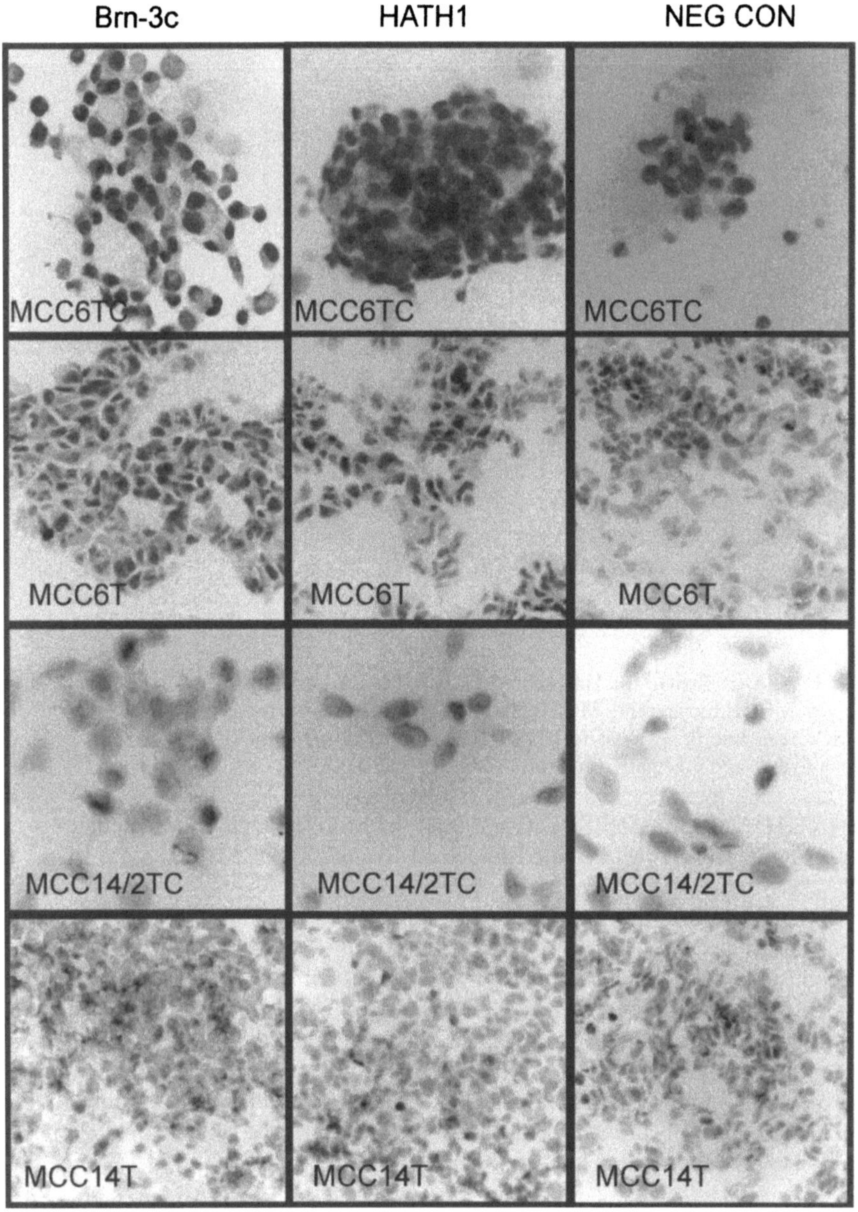

Fig. 5. Brn-3c, HATH1 and NSE reactivity in MCC biopsies and cell lines. MCC6 TC (Classic type I cell line) and MCC6 tumour have strong reactivity for Brn-3c, NSE and HATH1, MCC14/2 TC (Variant type IV cell line) and MCC14 biopsy are unreactive for NSE and HATH1 and show only non-specific cytoplasmic staining for Brn-3c, ×400

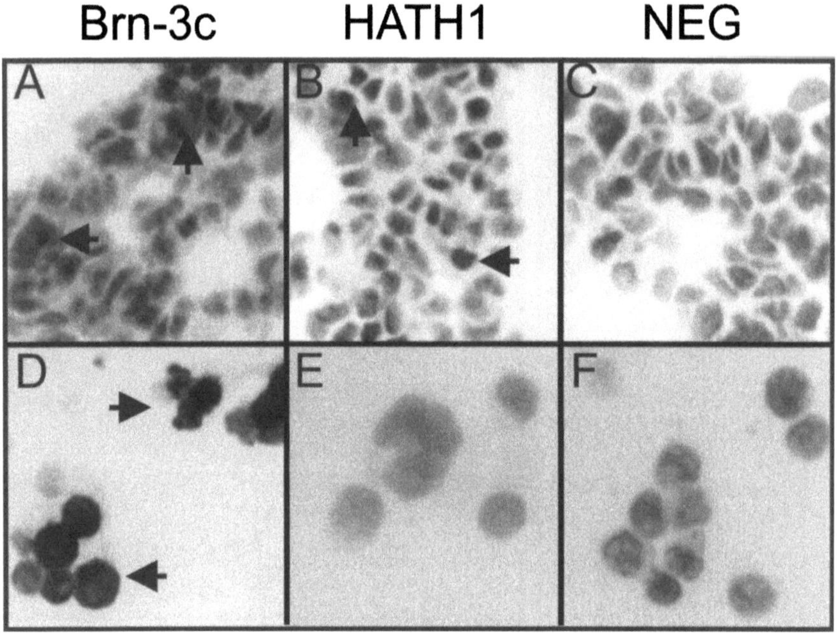

Fig. 6A–F. Brn-3c and HATH1 reactivity in MCC1 Variant suspension line
A MCC1 biopsy and MCC1cell line (**D**) are reactive for Brn-3c (*arrows*), but the MCC1cell line is negative for HATH1(**E**) although the biopsy (**B**) shows some cells are positive. **C** and **F** Negative controls, ×400

MATH1. Therefore, HES1 acts upstream of MATH1 with Brn-3c acting downstream of MATH1 (Muller and Littlewood-Evans 2001).

In order to determine if *HASH1* and *HES1* were also linked to neuroendocrine phenotype in MCC, RT-PCR with specific primers for *HASH1* and *HES1* was conducted. SCLC and neuroblastoma cell lines all had *HASH1* transcripts, but none of the MCC lines were positive (Fig. 4C). All cell lines tested had *HES1* transcripts and although the levels seen in MCC1and MCC6 were much lower then for the other lines (Fig. 4D), Western analysis revealed comparable levels of protein in all lines (Leonard et al. 2002). From these results, it seems unlikely that HASH1 controls neuroendocrine phenotype in MCC since neither Classic or Variant lines were positive.

Acknowledgements

This work was supported by an NH and MRC scholarship to J.H.L. and UQMYS scholarship and an AMRAD Student Travel Bursary to A.L.C. and by the Dept. of Oncology (formerly Queensland Radium Institute) Royal Brisbane

Hospital, the Royal Brisbane Hospital Research Into Cancer Trust Fund and the Queensland Cancer Fund. We would like to thank Drs M.M. Hughes, V.M. Hinkley, R.W. Allison, W. Cockburn, T.J. Harris and O. Williams for their support in collecting the specimens and Di Skilton, Registrar, Queensland Cancer Registry for MCC incidence information. We are grateful for the gifts of Brn-3a and Brn-3b (Dr Eric E. Turner, Department of Psychiatry and Program in Neuroscience, University of California, San Diego, USA) and MATH1 (Dr Jane E. Johnson, UT, Southwestern Medical Centre, Dallas, Texas, USA) antibodies.

References

Armstrong BK and Kricker A (2001) The epidemiology of UV-induced skin cancer. J Photochem Photobiol B Biol 63:8–18

Atanasoski S, Schreiber E, Fontana A, Herr W (1997) N-Oct 5 is generated by in vitro proteolysis of the neural POU-domain protein N-Oct 3. Oncogene 14:1287–1294

Ball DW, Azzoli CG, Baylin SB, Chi D, Dou S, Donis-Keller H, Cumaraswamy A, Borges M, Nelkin BD (1993) Identification of a human achaete-scute homolog highly expressed in neuroendocrine tumors. Proc Natl Acad Sci USA 90:5648–5652

Ben-Arie N, Hassan BA, Bermingham NA, Malicki DM, Armstrong D, Matzuk M, Bellen HJ, Zoghbi HY (2000) Functional conservation of atonal and Math1 in the CNS and PNS. Development 127:1039–1048

Bermingham NA, Hassan BA, Price SD, Vollrath MA, Ben-Arie N, Eatock RA, Bellen HJ, Lysakowski A, Zoghbi HY (1999) Math1: an essential gene for the generation of inner ear hair cells. Science 284:1837–1841

Budhram-Mahadeo V, Morris PJ, Smith MD, Midgley CA, Boxer LM, Latchman DS (1999a) p53 suppresses the activation of the Bcl-2 promoter by the Brn-3a POU family transcription factor. J Biol Chem 274:15237–15244

Budhram-Mahadeo V, Ndisang D, Ward T, Weber BL, Latchman DS (1999b) The Brn-3b POU family transcription factor represses expression of the BRCA-1 anti-oncogene in breast cancer cells. Oncogene 18:6684–6691

Carney DN, Gazdar AF, Bepler G, Guccion JG, Marangos PJ, Moody TW, Zweig MH, Minna JD (1985) Establishment and identification of small cell lung cancer cell lines having classic and variant features. Cancer Res 45:2913–2923

Chen H, Thiagalingam A, Chopra H, Borges MW, Feder JN, Nelkin BD, Baylin SB, Ball DW (1997) Conservation of the *Drosophila* lateral inhibition pathway in human lung cancer: a hairy-related protein (HES-1) directly represses achaete-scute homolog-1 expression. Proc Natl Acad Sci USA 94:5355–5360

Collum RG, Fisher PE, Datta M, Mellis S, Thiele C, Huebner K, Croce CM, Israel MA, Theil T, Moroy T (1992) A novel POU homeodomain gene specifically expressed in cells of the developing mammalian nervous system. Nucleic Acids Res 20:4919–4925

Compton CC, Regauer S, Seiler GR, Landry DB (1990) Human Merkel cell regeneration in skin derived from cultured keratinocyte grafts. Lab Invest 63:233–241

Eberl DF, Hardy RW, Kernan MJ (2000) Genetically similar transduction mechanisms for touch and hearing in *Drosophila*. J Neurosci 20:5981–5988

Erkman L, McEvilly RJ, Luo L, Ryan AK, Hooshmand F, O'Connell SM, Keithley EM, Rapaport DH, Ryan AF, Rosenfeld MG (1996) Role of transcription factors Brn-3.1 and Brn-3.2 in auditory and visual system development. Nature 381:603–606

Erkman L, Yates PA, McLaughlin T, McEvilly RJ, Whisenhunt T, O'Connell SM, Krones AI, Kirby MA, Rapaport DH, Bermingham JR, O'Leary DD, Rosenfeld MG (2000) A POU domain transcription factor-dependent program regulates axon pathfinding in the vertebrate visual system. Neuron 28:779–792

Gazdar AF, Carney DN, Nau MM, Minna JD (1985) Characterization of variant subclasses of cell lines derived from small cell lung cancer having distinctive biochemical, morphological, and growth properties. Cancer Res 45:2924–2930

Hara Y, Rovescalli AC, Kim Y, Nirenberg M (1992) Structure and evolution of four POU domain genes expressed in mouse brain. Proc Natl Acad Sci USA 89:3280–3284

Hartschuh W, Weihe E, Egner U (1989) Chromogranin A in the mammalian Merkel cell: cellular and subcellular distribution. J Invest Dermatol 93:641–648

He X, Treacy MN, Simmons DM, Ingraham HA, Swanson LW, Rosenfeld MG (1989) Expression of a large family of POU-domain regulatory genes in mammalian brain development. Nature 340:35–41

Helms AW, Abney AL, Ben-Arie N, Zoghbi HY, Johnson JE (2000) Autoregulation and multiple enhancers control Math1 expression in the developing nervous system. Development 127:1185–1196

Herr W, Cleary MA (1995) The POU domain: versatility in transcriptional regulation by a flexible two-in-one DNA-binding domain. Genes Dev 9:1679–1693

Herr W, Sturm RA, Clerc RG, Corcoran LM, Baltimore D, Sharp PA, Ingraham HA, Rosenfeld MG, Finney M, Ruvkun G (1988) The POU domain: a large conserved region in the mammalian pit-1, oct-1, oct-2, and *Caenorhabditis elegans* unc-86 gene products. Genes Dev 2:1513–1516

Huang EJ, Zang K, Schmidt A, Saulys A, Xiang M, Reichardt LF (1999) POU domain factor Brn-3a controls the differentiation and survival of trigeminal neurons by regulating Trk receptor expression. Development 126:2869–2882

Isaka F, Ishibashi M, Taki W, Hashimoto N, Nakanishi S, Kageyama R (1999) Ectopic expression of the bHLH gene Math1 disturbs neural development. Eur J Neurosci 11:2582–2588

Latchman DS (1999) POU family transcription factors in the nervous system. J Cell Physiol 179:126–133

Leblond-Francillard M, Picon A, Bertagna X, de Keyzer Y (1997) High expression of the POU factor Brn3a in aggressive neuroendocrine tumors. J Clin Endocrinol Metab 82:89–94

Leonard JH, Bell JR, Kearsley JH (1993) Characterization of cell lines established from Merkel-cell ("small-cell") carcinoma of the skin. Int J Cancer 55:803–810

Leonard JH, Dash P, Holland P, Kearsley JH, Bell JR (1995a) Characterisation of four Merkel cell carcinoma adherent cell lines. Int J Cancer 60:100–107

Leonard JH, Ramsay JR, Kearsley JH, Birrell GW (1995b) Radiation sensitivity of Merkel cell carcinoma cell lines. Int J Radiat Oncol Biol Phys 32:1401–1407

Leonard JH, Cook AL, Van Gele M, Boyle G, Inglis KJ, Speleman F, Sturm RA (2002) Proneoral and proendocrine transcription factor expression in cutaneous mechanoreceptors (Merkel) cell and Merkel cell carcinoma. Int J Cancer 102:103–110

Meeuwissen JA, Bourne RG, Kearsley JH (1995) The importance of postoperative radiation therapy in the treatment of Merkel cell carcinoma. Int J Radiat Oncol Biol Phys 31:325–331

Moll I, Kuhn C, Moll R (1995) Cytokeratin 20 is a general marker of cutaneous Merkel cells while certain neuronal proteins are absent. J Invest Dermatol 104:910–915

Moll I, Zieger W, Schmelz M (1996) Proliferative Merkel cells were not detected in human skin. Arch Dermatol Res 288:184–187

Monk M, Holding C (2001) Human embryonic genes re-expressed in cancer cells. Oncogene 20:8085–8091

Muller U, Littlewood-Evans A (2001) Mechanisms that regulate mechanosensory hair cell differentiation. Trends Cell Biol 11:334–342

Ndisdang D, Morris PJ, Chapman C, Ho L, Singer A, Latchman DS (1998) The HPV-activating cellular transcription factor Brn-3a is overexpressed in CIN3 cervical lesions. J Clin Invest 101:1687–1692

Popp S, Waltering S, Herbst C, Moll I, Boukamp P (2002) UV-B-type mutations and chromosomal imbalances indicate common pathways for the development of Merkel and skin squamous cell carcinomas. Int J Cancer 99:352–360

Przyborski SA, Morton IE, Wood A, Andrews PW (2000) Developmental regulation of neurogenesis in the pluripotent human embryonal carcinoma cell line NTERA-2. Eur J Neurosci 12:3521–3528

Rosenfeld MG (1991) POU-domain transcription factors: pou-er-ful developmental regulators. Genes Dev 5:897–907

Schöler HR, Hatzopoulos AK, Balling R, Suzuki N, Gruss P (1989) A family of octamer-specific proteins present during mouse embryogenesis: evidence for germline-specific expression of an Oct factor. EMBO J 8:2543–2550

Schreiber E, Himmelmann A, Malipiero U, Tobler A, Stahel R, Fontana A (1992) Human small cell lung cancer expresses the octamer DNA-binding and nervous system-specific transcription factor N-Oct-3 (brain-2). Cancer Res 52:6121–6124

Schreiber E, Merchant RE, Wiestler OD, Fontana A (1994) Primary brain tumors differ in their expression of octamer deoxyribonucleic acid-binding transcription factors from long-term cultured glioma cell lines. Neurosurgery 34:129–135

Siskind V, Aitken J, Green A, Martin N (2002) Sun exposure and interaction with family history in risk of melanoma, Queensland, Australia. Int J Cancer 97:90–95

Smith AG, Brightwell G, Smit SE, Parsons PG, Sturm RA (1998) Redox regulation of Brn-2/N-Oct-3 POU domain DNA binding activity and proteolytic formation of N-Oct-5 during melanoma cell nuclear extraction. Melanoma Res 8:2–10

Sturm RA, Herr W (1988) The POU domain is a bipartite DNA-binding structure. Nature 336:601–604

Sturm RA, Das G, Herr W (1988) The ubiquitous octamer-binding protein Oct-1 contains a POU domain with a homeo box subdomain. Genes Dev 2:1582–1599

Theil T, McLean-Hunter S, Zornig M, Moroy T (1993) Mouse Brn-3 family of POU transcription factors: a new aminoterminal domain is crucial for the oncogenic activity of Brn-3a. Nucleic Acids Res 21:5921–5929

Theil T, Rodel B, Spiegelhalter F, Moroy T (1995) Short isoform of POU factor Brn-3b can form a heterodimer with Brn-3a that is inactive for octamer motif binding. J Biol Chem 270:30958–30964

Thomson JA, Parsons PG, Sturm RA (1993) In vivo and in vitro expression of octamer binding proteins in human melanoma metastases, brain tissue, and fibroblasts. Pigment Cell Res 6:13–22

Thomson JA, Leonard JH, McGregor K, Sturm RA, Parsons PG (1994) A nonconsensus octamer-recognition sequence (TAATGARAT-motif) identifies a novel DNA binding protein in human Merkel cell carcinoma cell lines. Int J Cancer 58:285–290

Thomson JA, Murphy K, Baker E, Sutherland GR, Parsons PG, Sturm RA, Thomson F (1995) The brn-2 gene regulates the melanocytic phenotype and tumorigenic potential of human melanoma cells. Oncogene 11:691–700

Trieu M, Rhee JM, Fedtsova N, Turner EE (1999) Autoregulatory sequences are revealed by complex stability screening of the mouse brn-3.0 locus. J Neurosci 19:6549–6558

Vahava O, Morell R, Lynch ED, Weiss S, Kagan ME, Ahituv N, Morrow JE, Lee MK, Skvorak AB, Morton CC, Blumenfeld A, Frydman M, Friedman TB, King MC, Avraham KB (1998) Mutation in transcription factor POU4F3 associated with inherited progressive hearing loss in humans. Science 279:1950–1954

Van Gele M, Kaghad M, Leonard JH, Van Roy N, Naeyaert JM, Geerts ML, Van Belle S, Cocquyt V, Bridge J, Sciot R, De Wolf-Peeters C, De Paepe A, Caput D, Speleman F (2000) Mutation analysis of P73 and TP53 in Merkel cell carcinoma. Br J Cancer 82:23–826

Xiang M (1998) Requirement for Brn-3b in early differentiation of postmitotic retinal ganglion cell precursors. Dev Biol 197:55–169

Xiang M, Zhou L, Macke JP, Yoshioka T, Hendry SH, Eddy RL, Shows TB, Nathans J (1995) The Brn-3 family of POU-domain factors: primary structure, binding specificity, and expression in subsets of retinal ganglion cells and somatosensory neurons. J Neurosci 15:762–4785

Xiang M, Gan L, Li D, Chen ZY, Zhou L, O'Malley BW Jr, Klein W, Nathans J (1997) Essential role of POU-domain factor Brn-3c in auditory and vestibular hair cell development. Proc Natl Acad Sci USA 94:9445–9450

Xiang M, Gao WQ, Hasson T, Shin JJ (1998) Requirement for Brn-3c in maturation and survival, but not in fate determination of inner ear hair cells. Development 125:935–3946

Yang Q, Bermingham NA, Finegold MJ, Zoghbi HY (2001) Requirement of Math1 for secretory cell lineage commitment in the mouse intestine. Science 294:155–2158

Youl P, Aitken J, Hayward N, Hogg D, Liu L, Lassam N, Martin N, Green A (2002) Melanoma in adolescents: a case-control study of risk factors in Queensland, Australia. Int J Cancer 98:2–98

Expression Patterns of Connexins in Merkel Cell Carcinoma and Adjacent Epidermis

Nikolas K. Haass, Pia Houdek, Johanna M. Brandner and Ingrid Moll

Department of Dermatology and Venerology, University Hospital Hamburg-Eppendorf, Hamburg, Germany

Summary

Alterations in cell–cell communication play an important role in carcinogenesis. The specificity of communication is controlled by the type of connexins (Cx) forming gap junctions (GJ).
In this work, we show an induction of Cx30 and Cx26 in normal interfollicular epidermis adjacent to Merkel cell carcinoma. This observation and previous investigations of gap junctional communication in other malignant skin tumours give a hint that the change of the composition of gap junctions in the epidermis adjacent to malignant skin tumours may play a role in their metastasization.

Introduction

Gap junctions (GJ) play an important role in the communication of cells. They form channels between adjacent cells allowing the diffusion of molecules with a molecular weight of up to 1000 Da. These include ions, metabolites and second messenger molecules (Loewenstein 1981; Spray 1994). GJ consist of two hemichannels called connexons (one of each cell), each formed by six polypeptides belonging to a family of transmembrane proteins called connexins (Cx; for review see Beyer et al. 1990; Richard 2000). Various experiments have demonstrated the existence of homotypic and heterotypic channels, formed by identical or different connexons. Moreover, connexons can consist of either identical or different connexins (homomeric/heteromeric). However, not all connexins and connexons are compatible (Elfgang et al. 1995; Stauffer 1995; Jiang and Goodenough 1996). The type of connexins forming GJ influences the selectivity of the channels and thereby controls the specificity of cell–cell communication (Brissette et al. 1994; Elfgang et al. 1995; Veenstra 1996).

Ten of the known 21 different connexins are expressed in human skin. GJ have been supposed to be involved in the regulation of keratinocyte growth, differentiation and migration (Pitts et al. 1988; Salomon et al. 1993; Brissette et al. 1994; Goliger and Paul 1995). Moreover, Cx26 and Cx30 are induced during wound healing (Goliger and Paul 1995; Brandner, in prep.), psoriasis (Labarthe et al. 1998); and treatment with all-*trans*-retinoic-acid (Masgrau-Peya et al. 1997). Genetic studies have demonstrated the importance of epidermal gap junctions as mutations in their connexins are associated with autosomal dominant epidermal diseases.

There is evidence suggesting that connexins play a role in tumour biology (for review, see Krutovskikh and Yamasaki 1997; Li et al. 2002). Because of the shown changes of the expression pattern of Cx26, e.g., in hyperproliferative epidermis, we were interested in the distribution of Cx26 in skin tumours and their adjacent epidermis. We also investigated Cx30 because of its close relationship to the important, but until recently, not much understood Cx26. Because of the ubiquitous expression of Cx43 in normal skin, we were also interested in changes of its expression in skin tumours.

Recently, we have shown a change in the expression pattern of connexins in basal cell carcinoma, squamous cell carcinoma and malignant melanoma (Haass et al. 2001, 2002). There is not much known about the expression patterns of connexins in Merkel cell carcinoma and adjacent epidermis.

Material and Methods

Investigations were performed by immunofluorescence microscopy using previously described specific antibodies. Antibodies specific for Cx30 and Cx43 were purchased from Zymed Laboratories, San Francisco, CA, USA; the polyclonal antibody specific for Cx26 was made in our laboratory; the monoclonal antibody specific for CK20 was purchased from Progen Biotechnik, Heidelberg, Germany. For nuclear staining, DAPI (Boehringer Mannheim, Mannheim, Germany) was used.

Immunofluorescence microscopy was performed with an Axiophot II (Carl Zeiss, Jena/Oberkochen, Germany), photographed with a CCD Camera (Hamamatsu Photonics, Hamamatsu City, Japan) and documented using the software Openlab (Improvision, Lexington, MA, USA).

Results

As previously shown, both Cx30 and Cx26 are only weakly, if at all, expressed in normal interfollicular epidermis, while Cx43 is expressed ubiquitously in all cell layers.

Now we show that neither Cx30 (cf. Fig. 1a), Cx26 (cf. Fig. 1b) nor Cx43 (cf. Fig. 1c) occur in Merkel cell carcinoma. As expected, Cx43 is expressed throughout the interfollicular epidermis – adjacent to and also distant from the Merkel cell

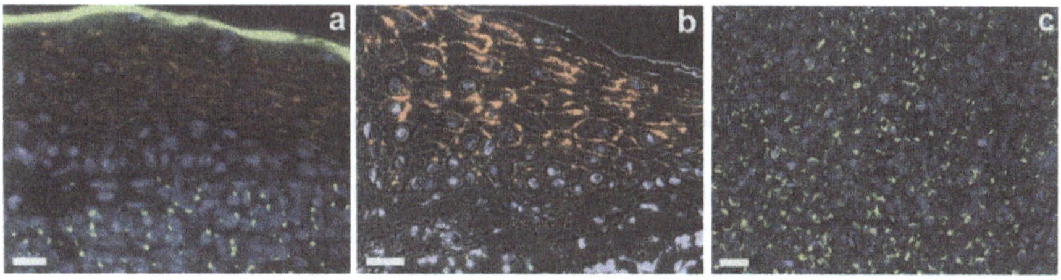

Fig. 1. a Immunofluorescence microscopy of a Merkel cell carcinoma and adjacent epidermis using antibodies against Cx30 (*red*) and CK20 (*green*), nuclear staining (DAPI, *blue*), overlaid with corresponding phase contrast. There is no expression of Cx30 in Merkel cell carcinoma cells, but note the expression of Cx30 in the epidermis adjacent to the Merkel cell carcinoma. *Bar* 6 μm. **b** Immunofluorescence microscopy of a Merkel cell carcinoma and adjacent epidermis using antibodies against Cx26 (*red*) and CK20 (*green*), nuclear staining (DAPI, *blue*), overlaid with corresponding phase contrast. There is no expression of Cx26 in Merkel cell carcinoma cells, but note the expression of Cx26 in the epidermis adjacent to the Merkel cell carcinoma. *Bar* 6 μm. **c** Immunofluorescence microscopy of a Merkel cell carcinoma using antibodies against Cx43 (*red*) and CK20 (*green*), nuclear staining (DAPI, *blue*), overlaid with corresponding phase contrast. There is no expression of Cx43 in Merkel cell carcinoma cells. *Bar* 6 μm

carcinoma (not shown). Moreover, as expected, there is no expression of Cx30 and Cx26 in unaffected interfollicular epidermis distant from the Merkel cell carcinoma (not shown). However, in striking contrast, both Cx30 (cf. Fig. 1a) and Cx26 (cf. Fig. 1b) are expressed in the interfollicular epidermis adjacent to Merkel cell carcinoma.

Discussion

These results clearly show an induction of Cx30 and Cx26 in normal interfollicular epidermis adjacent to Merkel cell carcinoma. Previously, we demonstrated similar results investigating gap junction proteins in malignant melanoma (Haass et al. 2002) and malignant epithelial skin tumours (Haass et al. 2001), but there was no such induction in the epidermis adjacent to semi-malignant (basal cell carcinoma) and benign lesions (melanocytic nevi). Therefore, one might suggest that the change of the composition of gap junctions in the epidermis adjacent to malignant skin tumours may play a role in the metastasation of these tumours.

References

Beyer EC, Paul DL, Goodenough DA (1990) Connexin family of gap junction proteins. J Membr Biol 116:187–194

Brissette JL, Kumar NM, Gilula NB, Hall JE, Dotto GP (1994) Switch in gap junction protein expression is associated with selective changes in junctional permeability during keratinocyte differentiation. Proc Natl Acad Sci USA 91:6453–6457

Elfgang C, Echert R, Lichtenbert-Frate H, Butterweck A, Traub O, Klein RA, Hülser DF, Willecke K (1995) Specific permeability and selective formation of gap junction channels in connexin-transfected HeLa cells. J Cell Biol 129:805–817

Goliger JA, Paul DL (1995) Wounding alters epidermal connexin expression and gap junction-mediated intercellular communication. Mol Biol Cell 6:1491–1501

Haass NK, Brandner JM, von den Driesch P, Kief S, Wladykowski E, Houdek P, Moll I (2001) Induction of gap junction proteins in epithelial skin tumours and adjacent epidermis. J Invest Dermatol 117:789 (Abstract)

Haass NK, Brandner JM, von den Driesch P, Wladykowski E, Kief S, Houdek P, Moll I (2002) Expression patterns of connexins in basal cell carcinoma, squamous cell carcinoma, malignant melanoma and adjacent epidermis. Arch Dermatol Res (Offprint) 294:72 (P229)

Jiang JX, Goodenough DA (1996) Heteromeric connexons in lens gap junction channels. Proc Natl Acad Sci USA 93:1287–1291

Krutovskikh V, Yamasaki H (1997) The role of gap junctional intercellular communication (GJIC) disorders in experimental and human carcinogenesis. Histol Histopathol 12:761–768

Labarthe MP, Bosco D, Saurat JH, Meda P, Salomon D (1998) Upregulation of Connexin 26 between keratinocytes of psoriatic lesions. J Invest Dermatol 111:72–76

Li G, Satayamoorthy K, Herlyn M (2002) Dynamics of cell interactions during melanoma development. Crit Rev Oral Biol Med 2002 13:62–70

Loewenstein WR (1981) Junctional intercellular communication: the cell-to-cell membrane channel. Physiol Rev 61:829–913

Masgrau-Peya E, Salomon D, Saurat JH, Meda P (1997) In vivo modulation of connexins 43 and 26 of human epidermis by topical retinoic acid treatment. J Histochem Cytochem 45:1207–1215

Pitts JD, Finbow ME, Kam E (1988) Junctional communication and cellular differentiation. Br J Cancer 9 (Suppl):52–57

Richard G (2000) Connexins: a connection with the skin. Exp Dermatol 9:77–96

Salomon D, Masgrau E, Vischer S, Chanson M, Saurat JH, Spray DS, Meda P (1993) Gap junction proteins and communication in human epidermis. Progr Cell Res 3:255–231

Spray DC (1994) Physiological and pharmacological regulation of gap junction channels. In: Citi S (ed) Molecular mechanisms of epithelial cell junctions: from development to disease. R G Landes Company, Austin, TX, pp 195–215

Stauffer KA (1995) The gap junction proteins β1-connexin (connexin-32) and β2-connexin (connexin26) can form heteromeric hemichannels. J Biol Chem 270:6768–6772

Veenstra RD (1996) Size and selectivity of gap junction channels formed from different connexins. J Bioenerg Biomembr 28:326–337

Expression Patterns of Tight Junction Proteins in Merkel Cell Carcinoma

Nikolas K. Haass, Pia Houdek, Ewa Wladykowski, Ingrid Moll and Johanna M. Brandner

Department of Dermatology and Venerology, University Hospital Hamburg-Eppendorf, Hamburg, Germany

Summary

Tight junctions (TJ) play a role in the barrier function of the epidermis. While Merkel cells are negative for Claudin 3, 4, 5, occludin and ZO-1, these TJ proteins are found in Merkel cell carcinoma. As TJ are able to connect neighbouring cells tightly and to control the paracellular pathway of substances, the formation of TJ might be a mechanism for a self-isolation of the tumour from its environment, especially from anti-tumorigenic cells and molecules. The synthesis of claudin 5, a marker for endothelial cells, in certain areas of Merkel cell carcinoma – in addition to blood vessels – gives a hint that a "vessel formation" might occur in this tumour.

Introduction

Despite the dogma that the barrier function of the epidermis is held upright only by the *stratum corneum*, i.e. corneocytes and lipids, it has recently been shown that cell–cell junctions, i.e. tight junctions and desmosomes, might very well be involved in this function too (i.e. Elias and Friend 1975; Elias et al. 1977, 2001; Morita et al. 1998; Pummi et al. 2001; Brandner et al. 2002; Furuse et al. 2002; Langbein et al. 2002). In simple epithelia and endothelia, tight junctions are responsible for the formation and maintenance of the tissue barrier between distinct compartments, e.g. the blood-brain barrier (for review, see Stevenson and Keon 1998). Tight junctions are composed of various transmembrane proteins [Claudin 1–20, Occludin, junctional adhesion molecule (JAM)] and plaque proteins (Zonula occludens proteins 1–3 (ZO-1–3), Symplekin).

As previously shown by us and others in normal interfollicular epidermis, Occludin and ZO-1 are expressed in the *stratum granulosum* and the transition layer, while Claudin 1 occurs in all layers (cf. Brandner et al. 2001, 2002).

There is evidence suggesting that tight junctions play a role in tumour biology (for review, see Martin and Jiang 2001). Recently, we have shown changes in expression patterns of tight junction proteins in malignant melanoma and adjacent epidermis compared to melanocytic nevi (Haass et al. 2002). There is not much known about the expression patterns of tight junction proteins in Merkel cell carcinoma.

Therefore, in this work, we focus on the occurrence of Claudin 1, 3, 4, and 5, Occludin and ZO-1 in Merkel cell carcinoma.

Material and Methods

Investigations were performed by immunofluorescence microscopy of frozen sections and of cultured cells using previously described specific antibodies.

Antibodies specific for Claudin 1, 3 and 5, Occludin and ZO-1 were purchased from Zymed Laboratories, San Francisco, CA, USA; the polyclonal antibody specific for Claudin 4 was made in our laboratory; the monoclonal antibody specific for CK20 was purchased from Progen Biotechnik, Heidelberg, Germany; the polyclonal antibody specific for CK20 was kindly provided from Prof. Dr. Roland Moll, Marburg, Germany. For nuclear staining, DAPI (Boehringer Mannheim, Mannheim, Germany) was used.

Immunofluorescence microscopy was performed with an Axiophot II (Carl Zeiss, Jena/Oberkochen, Germany), photographed with a CCD camera (Hamamatsu Photonics, Hamamatsu City, Japan) and documented using the software Openlab (Improvision, Lexington, MA, USA).

Results

	Merkel cell carcinoma	Remarks
Claudin 1 (Fig. 1a)	-	+ In adjacent epidermis
Claudin 3	+	
Claudin 4	+	
Claudin 5 (Fig. 1b)	Partially +	+ In blood vessels
Occludin (Fig. 1c)	+	
ZO-1	+	

Discussion

Claudin 3, 4, and 5, Occludin and ZO-1 were found in Merkel cell carcinoma cells. The expression of these molecules might result in functional tight junctions and, therefore, play a role in the isolation of the tumour from its environment.

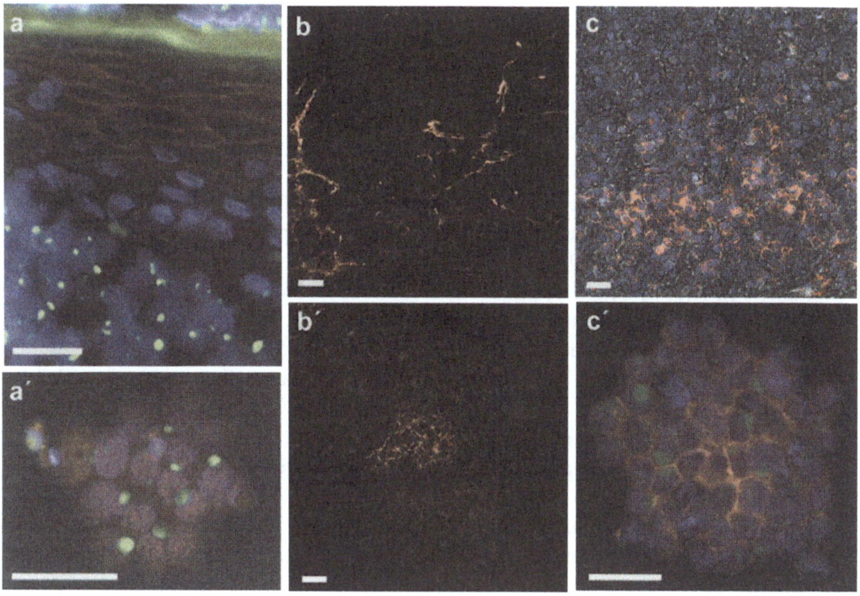

Fig. 1. a Immunofluorescence microscopy of a Merkel cell carcinoma and adjacent epidermis using antibodies against Claudin 1 (*red*) and CK20 (*green*), nuclear staining (DAPI, *blue*), overlaid with corresponding phase contrast. While there is the expected expression of Claudin 1 in the epidermis adjacent to the Merkel cell carcinoma, there is no expression of Claudin 1 in Merkel cell carcinoma cells. *Bar* 6 µm. **a'** Immunofluorescence microscopy of cultured Merkel cell carcinoma cells using antibodies against Claudin 1 (*red*) and CK20 (*green*), nuclear staining (DAPI, *blue*). Moreover, in cultured Merkel cell carcinoma cells there is no expression of Claudin 1. *Bar* 6 µm. **b** Immunofluorescence microscopy of a Merkel cell carcinoma using antibodies against Claudin 5 (*red*). There is the expected expression of Claudin 5 in blood vessels. *Bar* 6 µm. **b'** Immunofluorescence microscopy of a Merkel cell carcinoma using antibodies against Claudin 5 (*red*). Claudin 5 is also found in Merkel cell carcinoma cells. *Bar* 6 µm. **c** Immunofluorescence microscopy of a Merkel cell carcinoma using antibodies against Occludin (*red*), nuclear staining (DAPI, *blue*), overlaid with corresponding phase contrast. Occludin is expressed in Merkel cell carcinoma cells (as are Claudins 3, 4, 5 and ZO-1, not shown). *Bar* 6 µm. **c'** Immunofluorescence microscopy of cultured Merkel cell carcinoma cells using antibodies against Occludin (*red*) and CK20 (*green*), nuclear staining (DAPI, *blue*). Moreover, in cultured Merkel cell carcinoma cells Occludin is expressed at the cell borders in regions of confluency. *Bar* 6 µm

Claudin 1 is not found in Merkel cell carcinoma cells. This is a surprising result, because Claudin 1 is the only tight junction protein that appears in the environment of Merkel cells: the basal layer of the epidermis.

The endothelial marker Claudin 5 (Morita et al. 1999) is found in vessels, but also in Merkel cell carcinoma cells in certain parts of Merkel cell carcinoma. This result reminds one of the "vessel formation" by other malignant tumours such as malignant melanoma (i.e. Maniotis et al. 1999; Hendrix et al. 2002). Hence, this might be a hint that such a "vessel formation" occurs in Merkel cell carcinoma and

Claudin 5 is involved this process.

References

Brandner JM, Haass NK, Kief S, Wladikowski E, Houdek P, Moll I (2001) Expression and localisation of Tight-Junction-proteins in the human epidermis and various diseases. J Invest Dermatol 117:789

Brandner JM, Kief S, Grund C, Rendl M, Houdek P, Kuhn C, Tschachler E, Franke WW, Moll I (2002) Organization and formation of the tight junction-system in human epidermis and cultured keratinocytes. Eur J Cell Biol 81:253–263

Elias PM, Friend DS (1975) The permeability barrier in mammalian epidermis. J Cell Biol 65:180–191

Elias PM, McNutt NS, Friend DS (1977) Membrane alterations during cornification of mammalian squamous epithelia: a freeze-fracture, tracer, and thin-section study. Anat Rec 189:577–594

Elias PM, Matsuyoshi N, Wu H, Lin C, Wang ZH, Brown BE, Stanley JR (2001) Desmoglein isoform distribution affects stratum corneum structure and function. J Cell Biol 153:243–249

Furuse M, Hata M, Furuse K, Yoshida Y, Haratake A, Sugitani Y, Noda T, Kubo A, Tsukita S (2002) Claudin-based tight junctions are crucial for the mammalian epidermal barrier: a lesson from claudin-1-deficient mice. J Cell Biol 156:1099–1111

Haass NK, Brandner JM, von den Driesch P, Wladykowski E, Kief S, Houdek P, Moll I (2002) Changes in expression patterns of tight junction proteins in malignant melanoma and adjacent epidermis compared to melanocytic nevi. J Invest Dermatol 119:207 (Abstract)

Hendrix MJ, Seftor RE, Seftor EA, Gruman LM, Lee LM, Nickoloff BJ, Miele L, Sheriff DD, Schatteman GC (2002) Transendothelial function of human metastatic melanoma cells: role of the microenvironment in cell-fate determination. Cancer Res 62:665–668

Langbein L, Grund C, Kuhn C, Praetzel S, Kartenbeck J, Brandner JM, Moll I, Franke WW (2002) Tight junctions and compositionally related junctional structures in mammalian stratified epithelia and cell cultures derived therefrom. Eur J Cell Biol 81:419–435

Maniotis AJ, Folberg R, Hess A, Seftor EA, Gardner LM, Pe'er J, Trent JM, Meltzer PS, Hendrix MJ (1999) Vascular channel formation by human melanoma cells in vivo and in vitro: vasculogenic mimicry. Am J Pathol 155:739–752

Martin TA, Jiang WG (2001) Tight junctions and their role in cancer metastasis. Histol Histopathol 16:1183–1195

Morita K, Itoh M, Saitou M, Ando-Akatsuka Y, Furuse M, Yoneda K, Imamura S, Fujimoto K, Tsukita S (1998) Subcellular distribution of tight junction-associated proteins (occludin, ZO-1, ZO-2) in rodent skin. J Invest Dermatol 110:862–866

Morita K, Sasaki H, Furuse M, Tsukita S (1999) Endothelial claudin: claudin-5/TMVCF constitutes tight junction strands in endothelial cells. J Cell Biol 147:185–194

Pummi K, Malminen M, Aho H, Karvonen SL, Peltonen J, Peltonen S (2001) Epidermal tight junctions: ZO-1 and occludin are expressed in mature, developing, and affected skin and in vitro differentiating keratinocytes. J Invest Dermatol 117:1050–1058.

Stevenson BR, Keon BH (1998) The tight junction: morphology to molecules. Annu Rev Cell Dev Biol 14:89–109

Ganglioside Profiles of Merkel Cell Carcinoma

Ko Nagai, Hiroo Yasui, Toshiko Ishida, Gozo Muramoto, Chie Nakamura, Junji Nakano, Muto Masahiko

Department of Biomolecular Sensing (Dermatology), Yamaguchi University School of Medicine, Minami-Kogushi 1-1-1, Ube 755-8505, Japan

Summary

Merkel cells are located mainly in the epidermis and considered to have a regulatory role in the growth and differentiation of keratinocytes. On account of their biological behavior, Merkel cells are thought to belong to "paraneurons"(Fujita 1989). Although many kinds of functions of these cells have been clarified, their origin is still unknown. Therefore, we analyzed the ganglioside compositions of Merkel cell carcinomas in order to discuss their origin and metastatic potency.

Introduction

Gangliosides (sialylated glycosphingolipids) are components of mammalian plasma membranes and show quantitative and qualitative differences in their expression in different tissues. Many biological functions of gangliosides are known, such as growth regulation, cell differentiation, cell–substrate interaction, etc. Previously, we reported that GD3, a disialoganglioside, might play a role in the proliferation and metastasis of human malignant melanoma (Nakano 1996).

For a long time, Merkel cells have only been considered to be mechanoreceptors in the skin. However, many studies have revealed that they play a regulatory function in the development of epidermis and appendages during the fetal period (Narisawa 1993; Kim 1995). These behaviors resemble those of neuroendocrine cells. On the other hand, electron microscope studies suggest that Merkel cells might be derived from the epidermis (Munger 1965).

Merkel cell carcinoma is an uncommon malignant tumor which often occurs on the head and neck. This tumor sometimes metastases to lymph nodes or distant organs, but is considered to show better prognosis compared to malignant melanoma.

Patients and Control Materials

Two Japanese patients with Merkel cell carcinoma underwent excisional treatment in the Dermatology Clinic, Yamaguchi University Hospital. One was an 82-year-old female who had a dark red nodule on her right cheek. Another patient was a 58-year-old male who had a nodule on his chest. In both cases, the diagnosis was performed by routine pathological and immunohistochemical techniques. The specimens of Merkel cell carcinoma were stored at –80 °C until ganglioside analyses were performed. As controls, specimens obtained from nevus cell nevus, malignant melanoma, neurofibroma and dermatofibroma protuberans were also examined.

Isolation and Analysis of Gangliosides

Resected tumors were mashed and extracted with chloroform-methanol(2:1, 1:1, 1:2, vol/vol, sequentially). Then the glycolipid fraction was isolated by Florisil chromatography of acetylated derivatives. Neutral glycolipids were separated from gangliosides by DEAE-Sephadex A-50 ion exchange chromatography (Pharmacia /LKB, Piscataway, NJ). The ganglioside fraction was isolated by chromatography on a Sep-Pak C18 (Waters-Millipore, Milford, MA). Ganglioside composition was analyzed by thin-layer chromatography (TLC). TLC was performed in chloroform-methanol-0.2% CaCl (55:45:10) on high performance TLC plates (Silica Gel 60, Merck, Germany) and visualized with resorcinol-HCl. Plates were then scanned with a Shimadzu TLC scanner (SC930). As standard gangliosides, we used purified human brain gangliosides (GM2, GM1, GD2, GD3, GD1a, GD1b and GT1b) and purified canine erythrocytes (GM3).

Results

In Merkel cell carcinomas, we found several components of gangliosides; GM3, GD3, GD1a,GD1b and GT1b. The major component was GM3 amounting to 74.6% in case 1 and 55.7% in case 2. GD1b and GT1b were only present in traces or undetectable levels (0 and 0.5% in case 1, 0 and 1.2% in case 2). In neurofibromas and dermatofibrosarcoma protuberans, many kinds of gangliosides were found in addition to GM3 and GD3 (Table 1). In contrast to Merkel cell carcinomas, the major component of these two fibromatous tumors was GD3 or GT1b. Whereas the ganglioside profiles of nevus cell nevus were very similar to those of Merkel cell carcinomas, two cases of malignant melanomas (50% in examined cases) showed GD3 prominence compared to Merkel cell carcinomas.

Table 1. Composition(%) of gangliosides extracted from skin tumors.

	MCC1	MCC2	NCN1	NCN2	MM1	MM2	MM3	MM4	NF1	NF2	DFSP
GM3	74.6	55.7	62.4	61.6	84.2	41.2	25.6	75.1	31.0	17.3	25.7
GM2			1.1				4.6			0.9	8.8
GM1						2.4	2.1	8.3			
GD3	16.3	42.6	36.5	25.6	15.8	53.1	43.0	10.9	42.3	28.8	35.9
GD1a	9.1						12.5	2.3	6.7		
GD2							0.8			12.9	11.7
GD1b		0.5		3.8			3.3	3.4	5.5	5.7	4.8
GT1b		1.2		5.6			8.9		14.6	34.4	14.8
Others				3.4		3.3					

MCC: Merkel cell carcinoma; NCN: Nevus cell nevus; MM: Malignant melanoma; NF: Neurofibroma; DFSP: Dermatofibrosarcoma protuberans

Discussion

Merkel cell carcinoma is a malignant cutaneous tumor which often metastasizes to lymph nodes or distant organs. The origin of this tumor is still controversial. Until now, two hypotheses about the origin have been discussed; the neuroendocrine cell-origin theory and epidermal cell-origin theory.

Recent reports support the epidermal cell-origin theory on the basis of morphological and immunohistochemical data. In this study, we found the ganglioside profiles of Merkel cell carcinomas similar to those of nevus cell nevus. However, the ganglioside compositions of Merkel cell carcinomas were not always similar to those of malignant melanomas. In neurofibromas and dermatofibrosarcoma protuberans, we found very different profiles of gangliosides from those of Merkel cell carcinomas. These results may suggest a close relationship between Merkel cell carcinoma and nevus cell nevus. According to ganglioside profiles, we could not conclude that cells from Merkel cell carcinomas and cells from malignant melanomas had the same origin. However, cell function and malignant potency might affect the ganglioside compositions.

References

Fujita T (1989) Present status of paraneuron concept. Arch Histol Cytol 52(suppl):1–8

Kim DK (1995) The appearance, density, and distribution of Merkel cells inhuman embryonic and fetal skin: their relation to sweat gland and hair follicle development. J Invest Dermatol 104:411–416

Munger BL (1965) The intraepidermal innervation of the snout skin of the opossum. J Cell Biol 26:79–97

Nakano J (1996) Human melanoma cell lines deficient in GD3 ganglioside expression exhibit altered growth and tumorigenic characteristics. J Invest Dermatol 107:543–548

Narisawa Y (1993) Merkel cells in human fetal eccrine glands. Br J Dermatol 129:541–546

MOC-31, Cytokeratin 7 and S-100 Protein Immunoreactivity in Merkel Cell and Merkel Cell Carcinoma

T. García-Caballero[1], E. Pintos[2], R. Gallego[1], C. Parrado[3], M. Blanco[1], U.G. Falkmer[5], S. Falkmer[4], J. Forteza[2], and A. Beiras[1]

[1]Department of Morphological Sciences, University of Santiago de Compostela, Spain
[2]Department of Pathology, University of Santiago de Compostela, Spain
[3]Department of Histology and Pathology, University of Málaga, Spain
[4]Laboratory of Pathology and Clinical Cytology, Sant Olav's University Hospital., Trondheim, Norway
[5]Cancer Clinic, Sant Olav's University Hospital., Trondheim, Norway

Summary

The histological diagnosis of Merkel cell carcinoma can be difficult because it looks like other small blue cell tumors. In addition to the expected immunohistochemical results, some authors have reported reactivity for S-100 protein and CK 7 in Merkel cell carcinomas. The aim of the present work was to provide additional information about these unexpected immunoreactivities as well as to test the monoclonal antibody MOC-31 that was not previously described in Merkel cells or Merkel cell carcinomas. Nineteen cases of Merkel cell carcinoma were studied. MOC-31 plasma membrane immunoreactivity was found in 12 of 19 cases studied (63.1%). Immunoreactivity for CK 7 was observed in two cases (10.5%). S-100 protein was positive in four cases (21%). Normal human skin showed immunostaining for MOC-31 in the plasma membrane of virtually all Merkel cells. These cells were also immunoreactive for CK 7. S-100 protein was negative in human Merkel cells, but intensely positive in pig snout Merkel cells. In conclusion, normal human Merkel cells showed MOC-31 and CK 7 immunostaining and positivity for MOC-31, CK 7, and S-100 protein do not exclude the diagnosis of Merkel cell carcinoma.

Introduction

The histological diagnosis of Merkel cell carcinoma can be difficult because it looks like other small blue cell tumors. Immunohistochemical positivity for cytokeratin (CK) 20, neuroendocrine markers and neurofilaments confirm the diagnosis. In addition to these markers, some authors have reported reactivity for S-100 protein (Voigt et al. 1985; Drijkoningen et al. 1986; Heenan et al. 1990; Skoog et al. 1990; Kontochristopoulos et al. 2000) and CK 7 (Jensen et al. 2000) in Merkel cell carcinomas. The aim of the present work was to provide additional information about these unexpected immunoreactivities as well as to test the monoclonal antibody MOC-31 in normal Merkel cells and Merkel cell carcinomas. This antibody recognises the human pancarcinoma-associated epithelial glycoprotein-2 (EGP-2), also known as Ep-CAM, a 38-kDa transmembrane glycoprotein. MOC-31 has been assigned to the SCLC (small cell lung cancer)-cluster 2 of antibodies (Souhami et al. 1991) and it was demonstrated to be also useful in the differential diagnosis between adenocarcinoma and mesothelioma (Edwards and Oates 1995; Sosolik et al. 1997; Ordoñez 1998), but to our knowledge it was not previously described in Merkel cells or Merkel cell carcinomas.

Materials and Methods

Nineteen cases of Merkel cell carcinoma were obtained from the files of the Clinical University Hospital of Santiago de Compostela, Spain (9 cases) and the Department of Oncology and Pathology, Karolinska Hospital, Stockholm, Sweden (10 cases). Thirteen patients were female and six male, and patient age ranged from 20 to 93 (mean, 70.6 years). The most common localization was extremities (n=9) and head (n=8). Immunohistochemistry for CK 20 (Dako), chromogranin (Biogenex), MOC-31 (Dako), CK 7 (Dako) and S-100 protein (Dako) was automatically performed (TechMate 500) using the DAKO Envision staining procedure and diaminobenzidine as chromogen.

Results

By immunohistochemical techniques all tumors but three were positive for CK 20 and all cases but one were positive for chromogranin. MOC-31 plasma membrane immunoreactivity was found in 12 of 19 cases studied (63.1%). Most of them showed a diffuse pattern, but local immunoreactivity was also found (Fig. 1A, B). Immunoreactivity for CK 7 was observed in two cases (10.5%). One of them showed small and pleomorphic giant cells, both cell types being positive for CK 7 and CK 20 (Fig. 2A, B). S-100 protein was positive in four cases (21%) with nuclear and cytoplasmic immunostaining (Fig. 3A, B).

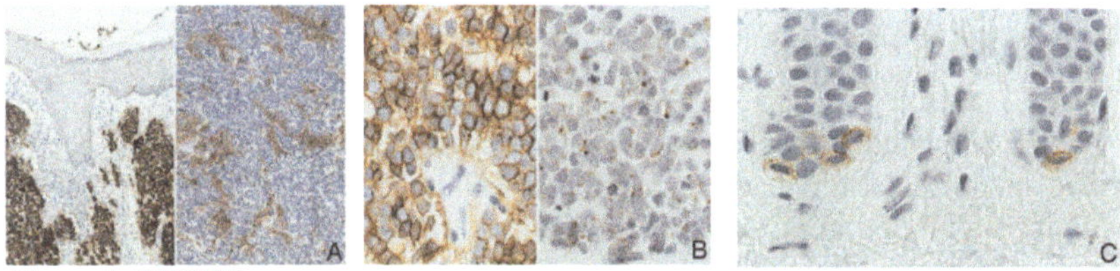

Figure 1: MOC-31

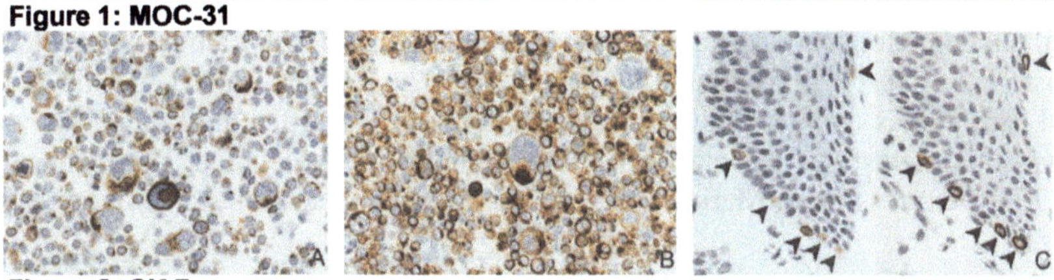

Figure 2: CK 7

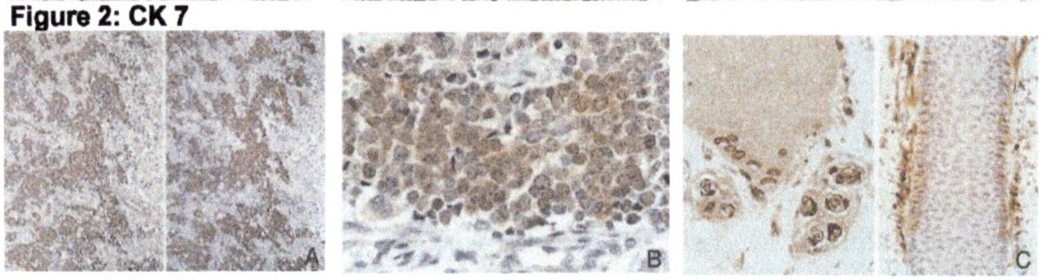

Figure 3: S-100 Protein

Fig. 1A–C. MOC-31. **A** Merkel cell carcinomas showed diffuse (*left*) or focal (*right*) immunoreactivity for MOC-31 (×10). **B** At higher magnification it was demonstrated that MOC-31 presented plasma membrane immunostaining (*left*), whereas CK 20 showed a dot-like pattern (*right*) (×60). **C** Normal Merkel cells of human fingertip also displayed MOC-31 immunoreactivity in the plasma membrane (×100)

Fig. 2A–C. CK 7. **A, B** Case of Merkel cell carcinoma positive for CK 7 (**A**) and CK 20 (**B**). Immunoreactivity was found in both small cells and giant pleomorphic cells (×40). **C** Normal human Merkel cells (*arrowheads*) were positive for CK 7 (*left*), but the intensity of immunostaining was weaker than for CK 20 (*right*) (×60)

Fig. 3A–C. S-100 PROTEIN. **A** Serial sections of a Merkel cell carcinoma showing immunoreactivity for S-100 protein (*left*) and CK 20 (*right*) (×10). **B** Higher magnification evidenced a nuclear and cytoplasmic positivity for S-100 protein (×40). **C** Intense immunostaining for S-100 protein was found in pig snout skin Merkel cells situated in both epidermis (*left*) and vibrissae (*right*) (×40)

The same markers were studied in normal Merkel cells of human fingertip and pig snout skin. Normal human skin showed immunostaining for MOC-31 in the plasma membrane of virtually all Merkel cells (Fig. 1C). These cells were also immunoreactive for CK 7, although with lower intensity than for CK 20 (Fig. 2C). S-100 protein was negative in human Merkel cells, but intensely positive in pig snout Merkel cells, with nuclear and cytoplasmic positivity (Fig. 3C).

Discussion

To our knowledge, MOC-31 immunoreactivity has not previously been described in normal Merkel cells or Merkel cell carcinomas. However, MOC-31 positivity was found in most carcinomas and is frequently used in the immunohistochemical battery for small cell carcinomas; it has also been proposed for radioimmunodetection and immunotherapy of these tumors (Myklebust et al. 1993; Kosterink et al. 1995; McLaughlin et al. 1999). The finding of MOC-31 in Merkel cell carcinomas excludes the use of this marker in the differential diagnosis with cutaneous metastases of small cell lung carcinomas. The clinical significance of MOC-31 expression in Merkel cell carcinomas remains to be studied in larger series.

Marked expression of CK 20 without concomitant CK 7 was assumed to be typical for both colorectal adenocarcinoma and Merkel cell carcinoma, and CK 7, which identifies bronchial small-cell carcinoma, is normally used for the differential diagnosis with Merkel cell carcinoma (Goessling et al. 2002). Our finding of CK 7 in occasional Merkel cell carcinomas confirms the results reported by Jensen et al. (2000). Normal Merkel cells also showed CK 7 immunoreactivity. This result agrees with a previous paper by Lundquist et al. (1999). These authors reported that Merkel cells express CK 7 with lower frequency than CK 20. However, we found that virtually all Merkel cells were positive for CK 7, although some of them showed very weak immunostaining. The lower frequency reported could be explained assuming that Merkel cells express CK 7 in smaller quantities than CK 20. The sensitivity of the immunohistochemical technique may not always be sufficient to detect all Merkel cells by CK 7.

In cutaneous pathology, S-100 protein is primarily used as a marker for the diagnosis of malignant melanoma. However, we showed occasional immunoreactivity for S-100 in Merkel cell carcinomas as was previously described by several authors (Voigt et al. 1985; Drijkoningen et al. 1986; Heenan et al. 1990; Skoog et al. 1990; Kontochristopoulos et al. 2000). S-100 protein expression in normal Merkel cells shows important differences between species. We found S-100 protein immunoreactivity in Merkel cells of pig snout skin, as previously described by Hartschuh and Weihe (1988). However, we did not find immunostaining for S-100 protein in human Merkel cells and the same result was reported by Moll et al. (1996) in rodents.

In conclusion, (1) normal human Merkel cells showed MOC-31, CK 7 immunostaining and S-100 protein was only found in pig snout Merkel cells, and (2) positivity for MOC-31, CK 7 and S-100 protein does not exclude the diagnosis of Merkel cell carcinoma.

References

Drijkoningen M, de Wolf-Peeters C, van Limbergen E, Desmet V (1986) Merkel cell tumor of the skin: an immunohistochemical study. Hum Pathol 17:301–307

Edwards C, Oates J (1995) OV632 and MOC-31 in the diagnosis of mesothelioma and adenocarcinoma: an assessment of their use in formalin-fixed and paraffin-wax embedded material. J Clin Pathol 48:626–630

Goessling W, McKee PH, Mayer RJ (2002) Merkel cell carcinoma. J Clin Oncol 20:588–598

Hartschuh W, Weihe E (1988) Multiple messenger candidates and marker substances in the mammalian Merkel cell-axon complex: a light and electron microscopic immunohistochemical study. In: Hamann W, Iggo A (eds) Progress in brain research, vol 74. Elsevier, Amsterdam, pp 181–187

Heenan PJ, Cole JM, Spagnolo DV (1990) Primary cutaneous neuroendocrine carcinoma (Merkel cell tumor). An adnexal epithelial neoplasm. Am J Dermatopathol 12:7–16

Jensen K, Kohler S, Rouse RV (2000) Cytokeratin staining in Merkel cell carcinoma: an immunohistochemical study of cytokeratins 5/6, 7, 17 and 20. Appl Immunohistochem Mol Morphol 8:310–315

Kontochristopoulos GJ, Stavropoulos PG, Krasagakis K, Goerdt S, Zouboulis CC (2000) Differentiation between Merkel cell carcinoma and malignant melanoma: an immunohistochemical study. Dermatology 201:123–126

Kosterink JG, de Jonge MW, Smit EF, Piers DA, Kengen RA, Postmus PE, Shochat D, Groen HJ, The HT, de Leij L (1995) Pharmacokinetics and scintigraphy of indium-111-DTPA-MOC-31 in small-cell lung carcinoma. J Nucl Med 36:2356–2362

Lundquist K, Kohler S, Rouse RV (1999) Intraepidermal cytokeratin 7 expression is not restricted to Paget cells, but is also seen in Toker cells and Merkel cells. Am J Surg Pathol 23:212–219

McLaughlin PM, Kroesen BJ, Dokter WH, van der Molen H, de Groot M, Brinker MG, Kok K, Ruiters MH, Buys CH, de Leij LF (1999) An EGP-2/Ep-CAM-expressing transgenic rat model to evaluate antibody-mediated immunotherapy. Cancer Immunol Immunother 48:303–311

Moll I, Paus R, Moll R (1996) Merkel cells in mouse skin: intermediate filament pattern, localization, and hair cycle-dependent density. J Invest Dermatol 106:281–286

Myklebust AT, Godal A, Pharo A, Juell S, Fodstad O (1993) Eradication of small cell lung cancer cells from human bone marrow with immunotoxins. Cancer Res 53:3784–3788

Ordoñez NG (1998) Value of the MOC-31 monoclonal antibody in differentiating epithelial pleural mesothelioma from lung adenocarcinoma. Hum Pathol 29:166–169

Skoog L, Schmitt FC, Tani E (1990) Neuroendocrine (Merkel-cell) carcinoma of the skin: immunocytochemical and cytomorphologic analysis of fine-needle aspirates. Diagn Cytopathol 6:53–57

Sosolik RC, McGaughy VR, De Young BR (1997) Anti-MOC31. A potential addition to the pulmonary adenocarcinoma versus mesothelioma immunohistochemistry panel. Mod Pathol 10:716–719

Souhami RL, Beverley PC, Bobrow LG, Ledermann JA (1991) Antigens of lung cancer: results of the Second International Workshop on Lung Cancer Antigens. J Natl Cancer Inst 83:609–612

Voigt JJ, Alsaati T, Gorguet B, Caveriviere P, Scarna H, Bugat R, Delsol G (1985) Carcinome à cellules de Merkel de la peau. Etude anatomo-clinique, ultrastructurale et immunohistochimique de 14 cas. Ann Pathol 5:195–203

Merkel Cell Carcinomas Possess Telomerase Activity

H. Stöppler[1], M. Conrad Stöppler[2], M. Kisiela[1], A. Holzbach[2], I. Moll[3], P. Houdek[3], and R. Moll[2]

[1]Department of Pharmacology and Toxicology, Philipps University of Marburg, 35033 Marburg, Germany

[2]Institute of Pathology, Philipps University of Marburg, 35033 Marburg, Germany

[3]Department of Dermatology and Venerology, University Hospital Hamburg-Eppendorf, 20246 Hamburg, Germany

Summary

Merkel cell carcinomas are believed to be derived from the Merkel cells of the epidermis and hair follicles. The histogenesis of the Merkel cells has not yet been fully elucidated, but several studies suggest that they might originate from an asymmetric cell division of basal keratinocytes or epithelial stem cells of the fetal epidermis, and the resulting differentiated Merkel cells have presumably lost their growth potential.

The capability of indefinite cell division in germ line cells and in the great majority of malignant tumors as well as an increased growth potential in certain somatic cells (such as basal cells of renewable tissues) is correlated with cellular telomerase activity, which is absent in differentiated somatic cells. In this study, the telomerase activity in cryostat sections of frozen Merkel cell tumor biopsies and in in vitro cultivated Merkel cell carcinoma cells was analyzed. The presence of the typical Merkel cell carcinoma phenotype was confirmed in all tumor specimens and cell lines by immunohistochemical staining for the cytokeratin CK 20, which was present in the typical plaque-like distribution. Microdissection was employed on the cryostat sections to concentrate tumor material and to remove potentially telomerase-positive contaminating elements such as epidermis and hair follicles. Telomerase activity was measured by the polymerase chain reaction-based telomeric repeat amplification protocol (TRAP) assay.

We detected telomerase activity in all four tumors studied and three of four cell lines. Control experiments showed a lack of telomerase activity in normal liver and kidney tissue and telomerase positivity in a breast carcinoma biopsy and activated lymph node, as expected. Immunohistochemistry demonstrated the presence

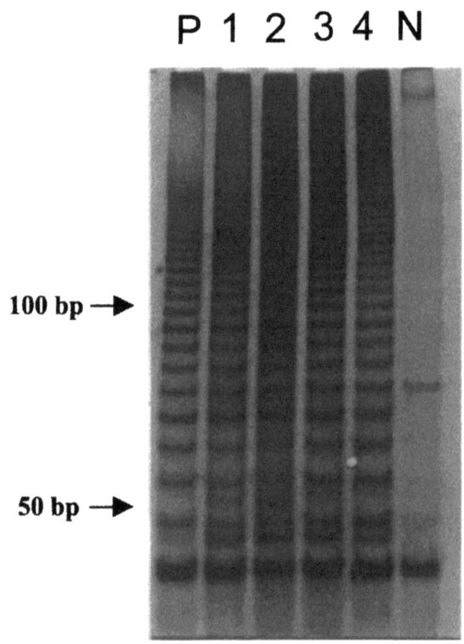

Fig. 1. Telomerase activity of Merkel cell carcinoma tissues. Depicted are the TRAP assay products following gel electrophoresis for a positive control cell line (*P*), four Merkel cell carcinoma tumor specimens (*1–4*) and normal liver tissue as a negative control (*N*). The presence of the characteristic "ladder" of DNA fragments indicative of a positive TRAP assay is seen in the positive control lane and in all four tumor samples

of the catalytic subunit of telomerase in nearly all nuclei of Merkel cell carcinomas.

These results show that despite their pronounced neuroendocrine differentiation and their occurrence in patients of advanced age, Merkel cell carcinomas possess telomerase activity similar to that of common carcinoma types. In future studies, the (difficult) procurement and subsequent analysis of early stages of Merkel cell carcinomas might help to interpret the present findings and to reveal more about the biology of this peculiar type of malignant tumor.

Methods and Results

Immunohistochemical Characterization of Merkel Cell Carcinomas by Staining for Cytokeratins and Leukocyte Common Antigen

Cryostat sections of frozen biopsies from the four Merkel cell tumors studied were cut and microscopically evaluated after hematoxylin and eosin staining to verify the presence of tumor tissue. Immunohistochemical staining confirmed the

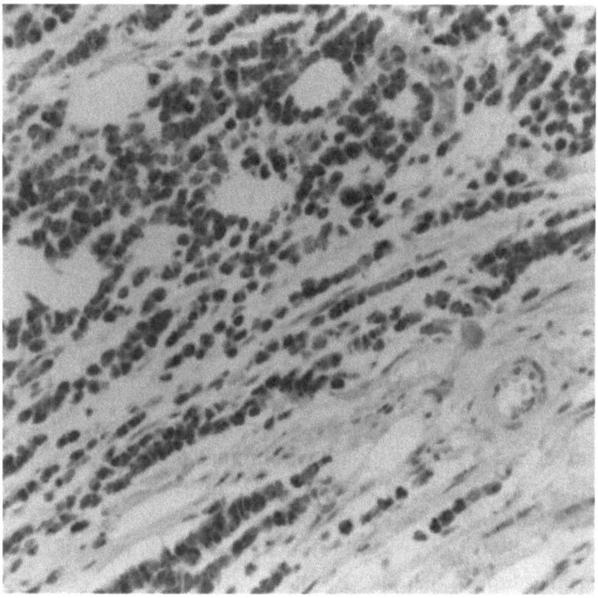

Fig. 2. Immunohistochemical staining for the catalytic subunit of telomerase. Merkel cell carcinoma sections were stained with a monoclonal antibody directed against the catalytic subunit of the telomerase enzyme (NCL-hTERT, clone 44F12, Novocastra Laboratories, Newcastle/Medac, Wedel, Germany). Virtually 100% of the tumor cells showed a characteristic strong nuclear staining, while adjoining connective tissue elements were negative. Magnification ×170

diagnosis of the tumor tissue as Merkel cell carcinomas, demonstrating strong and typically plaque-like reactions for CK 20. In the tumor stroma, variable proportions of small (nonactivated) lymphocytes were present as verified by immunostaining for leukocyte common antigen (LCA, CD45). The relative proportion of lymphocytes within the tumor tissues was estimated in each case since infiltrating lymphocytes could influence the TRAP assays. These proportions ranged from sparse (~1 per 100 tumor cells) to moderately abundant (~1 per 4 tumor cells) lymphocytes. The Merkel cell carcinoma cell cultures also showed typical plaque-like CK 20 staining.

Analysis of Telomerase Activity of Merkel Cell Carcinoma Biopsies and in Vitro Established Merkel Cell Carcinoma Cultures by the Telomeric Repeat Amplification Protocol (TRAP) Assay

Cryostat sections of four frozen Merkel cell carcinoma biopsies and cells of in vitro established Merkel cell carcinoma cultures were lysed, and 1 μg of cytoplasmic protein extract was tested in a telomeric repeat amplification protocol (TRAP) assay for telomerase activity. Thirty percent of the TRAP-assay products were separated on an 8% nondenaturing polyacrylamide gel, which was then stained with silver (Fig. 1).

Immunohistochemical Staining for the catalytic Subunit of the Telomerase Enzyme

Immunohistochemical staining of Merkel cell tumors with a monoclonal antibody directed against the catalytic subunit of telomerase showed a homogeneous nuclear staining pattern in nearly all of the tumor cells (Fig. 2) while nuclei of adjacent stromal cells were negative.

Conclusions

Despite their pronounced neuroendocrine differentiation and their occurrence in patients of advanced age, Merkel cell carcinomas possess telomerase activity similar to that of common carcinoma types. Since three of the four tumors analyzed in this study were microdissected prior to analysis, we conclude that microdissection is an effective method for removal of surrounding and possibly contaminating tissue in the preparation of samples for assays of telomerase activity. The microdissection procedure itself, which theoretically could lead to release of chromosomal DNA and false-positive TRAP assay results, does not interfere with the TRAP assay for telomerase activity.

References

Greider CW (1998) Telomerase activity, cell proliferation, and cancer. Proc Natl Acad Sci USA 95:90–92

Harley CB, Futcher AB, Greider CW (1990) Telomeres shorten during ageing of human fibroblasts. Nature 345:458–460

Hiyama E, Hiyama K, Yokoyama T et al. (2001) Immunohistochemical detection of telomerase (hTERT) protein in human cancer tissues and a subset of cells in normal tissue. Neoplasia 3:17–26

Liu K, Schoonmaker MM, Levine BL, June CH, Hodes RJ, Weng NP (1999) Constitutive and regulated expression of telomerase reverse transcriptase (hTERT) in human lymphocytes. Proc Natl Acad Sci USA 96:5147–5152

Stöppler H, Conrad Stöppler M, Holzbach A, Houdek P, Moll I, Moll R (2001) Telomerase activity of Merkel cell carcinomas and Merkel cell carcinoma-derived cell lines. Arch Dermatol Res 293:397–406

Chemotherapy in Merkel Cell Carcinoma with Distant Metastasis – Proposal of a Pan-European Clinical Trial

Peter Helmbold[1] and Axel Grothey[2]

[1]Department of Dermatology, Martin-Luther-University Halle-Wittenberg, Halle (Saale), Germany

[2]Department of Medicine IV, Martin-Luther-University Halle-Wittenberg, Halle (Saale), Germany

Merkel cell carcinoma (MCC) is a rare and highly malignant neuroendocrine skin tumour characterized by a high frequency of metastases (Helmbold et al. 2001). The 5-year survival rate is stage-dependent and ranges from 30 to 74% (Morrison et al. 1990; Meeuwissen et al. 1995; Skelton et al. 1997; Allen et al. 1999). While the combination of surgery and subsequent adjuvant radiotherapy appears to provide optimal local control in stages I (primary) and II (local metastasis), the therapy of stage III (distant metastasis) remains to be defined. The prognosis in stage III MCC is poor. Chemotherapy appears to be the most successful approach in stage III MCC. Nevertheless, a standard chemotherapy for MCC remains to be identified. A variety of different chemotherapy protocols have been used in the treatment of advanced MCC. However, even the few larger studies in this field have the character of extended case reports. Currently, there is *no ongoing trial on chemotherapy or other systemic drugs in MCC.* Experience is available for cytostatic drugs like cyclophosphamide, doxorubicin, epirubicin, vincristine, etoposide, cisplatin, carboplatin, 5-fluorouracil, dacarbazine, mitoxantrone, bleomycin, and ifosfamide (Voog et al. 1999; Tai et al. 2000).

To identify the optimal chemotherapy in metastatic MCC, the following aspects were taken into account for planning a Pan-European study:

- MCC patients are mostly older persons, who tend to tolerate aggressive chemotherapy poorly.
- The most successful agents used in MCC chemotherapy were cisplatin derivates and etoposide.
- Etoposide can be given as oral monotherapy. Based on the literature (Fenig et al. 2000) and own experience (P. Helmbold), this therapy might be equivalent to more aggressive therapies and can be continued for a longer time (disease control).
- Cisplatin can be substituted by the better tolerable carboplatin.

- Recently, Greco et al. (2001) has reported on taxane-based combination chemotherapy (paclitaxel/carboplatin/etoposid) for patients with carcinoma of unknown primary site. They found long-term complete remissions only in those cases with small cell neuroendocrine-like tumours.
- Because of the known chemosensitivity, a control arm without treatment should not be included in a study on chemotherapy in MCC.

These points enforced us to suggest a comparison of the following two protocols in a Pan-European Multicentre randomised phase II study:

Therapy and Objective

Paclitaxel – Carboplatin – Etoposide (TCE) (Greco et al. 2001)

Paclitaxel 200 mg/m^2 intravenously (i.v.) d 1, Carboplatin AUC=6 i.v. d 1, and oral VP16 50 mg and 100 mg alternating, d 1–10, q d 22 versus
Oral Etoposide monotherapy: VP16 2×50 mg, d 1–14, q d 22)
Primary objective: Progression free survival (PFS)
Secondary objectives: Response rate, overall survival, toxicity, quality of life

Timetable

- Recruitment of patients: 3 years
- Primary objective (PSF) +1 year
- Maximal individual follow-up +5 years

For further information please contact:

PD Dr. Peter Helmbold
Department of Dermatology
Martin-Luther-University Halle-Wittenberg
Ernst-Kromayer-Str. 5–6
06097 Halle (Saale), Germany
Tel.: +49-(0)345-5573911, Fax: +49-(0)345-5573941
e-mail: peter.helmbold@medizin.uni-halle.de

References

Allen PJ, Zhang ZF, Coit DG (1999) Surgical management of Merkel cell carcinoma. Ann Surg 229:97–105

Fenig E, Brenner B, Njuguna E, Katz A, Schachter J, Sulkes A (2000) Oral etoposide for Merkel cell carcinoma in patients previously treated with intravenous etoposide. Am J Clin Oncol 23:65–67

Greco FA, Gray J, Burris HA, 3rd, Erland JB, Morrissey LH, Hainsworth JD (2001) Taxane-based chemotherapy for patients with carcinoma of unknown primary site. Cancer J 7:203–212

Helmbold P, Schröter S, Holzhausen HJ, Hartschuh W, Marsch WC (2001). Das Merkelzellcarcinom: eine diagnostische und therapeutische Herausforderung. Chirurgie 72:396–401

Meeuwissen JA, Bourne RG, Kearsley JH (1995) The importance of postoperative radiation therapy in the treatment of Merkel cell carcinoma. Int J Radiat Oncol Biol Phys 31:325–331

Morrison WH, Peters LJ, Silva EG, Wendt CD, Ang KK, Goepfert H (1990) The essential role of radiation therapy in securing locoregional control of Merkel cell carcinoma. Int J Radiat Oncol Biol Phys 19:583–591

Skelton HG, Smith KJ, Hitchcock CL, McCarthy WF, Lupton GP, Graham JH (1997) Merkel cell carcinoma: analysis of clinical, histologic, and immunohistologic features of 132 cases with relation to survival. J Am Acad Dermatol 37:734–739

Tai PT, Yu E, Winquist E, Hammond A, Stitt L, Tonita J, Gilchrist J (2000) Chemotherapy in neuroendocrine/Merkel cell carcinoma of the skin: case series and review of 204 cases. J Clin Oncol 18:2493–2499

Voog E, Biron P, Martin JP, Blay JY (1999) Chemotherapy for patients with locally advanced or metastatic Merkel cell carcinoma. Cancer 85:2589–2595

Authors Index

Subject Index

MIX
Papier aus verantwortungsvollen Quellen
Paper from responsible sources
FSC® C105338
FSC
www.fsc.org

If you have any concerns about our products,
you can contact us on
ProductSafety@springernature.com

In case Publisher is established outside the EU,
the EU authorized representative is:
Springer Nature Customer Service Center GmbH
Europaplatz 3, 69115 Heidelberg, Germany

Printed by Libri Plureos GmbH
in Hamburg, Germany